Springer Series in Advanced Manufacturing

Series Editor

Duc Truong Pham, University of Birmingham, Birmingham, UK

The **Springer Series in Advanced Manufacturing** includes advanced textbooks, research monographs, edited works and conference proceedings covering all major subjects in the field of advanced manufacturing.
The following is a non-exclusive list of subjects relevant to the series:

1. Manufacturing processes and operations (material processing; assembly; test and inspection; packaging and shipping).
2. Manufacturing product and process design (product design; product data management; product development; manufacturing system planning).
3. Enterprise management (product life cycle management; production planning and control; quality management).

Emphasis will be placed on novel material of topical interest (for example, books on nanomanufacturing) as well as new treatments of more traditional areas.

As advanced manufacturing usually involves extensive use of information and communication technology (ICT), books dealing with advanced ICT tools for advanced manufacturing are also of interest to the Series.

Springer and Professor Pham welcome book ideas from authors. Potential authors who wish to submit a book proposal should contact Anthony Doyle, Executive Editor, Springer, e-mail: anthony.doyle@springer.com.

More information about this series at http://www.springer.com/series/7113

Wit Grzesik · Adam Ruszaj

Hybrid Manufacturing Processes

Physical Fundamentals, Modelling and Rational Applications

Springer

Wit Grzesik
Department of Manufacturing
and Materials Engineering
Opole University of Technology
Opole, Poland

Adam Ruszaj
Faculty of Mechanical Engineering
Cracow University of Technology
Kraków, Poland

State University of Applied Sciences
Nowy Sącz, Poland

ISSN 1860-5168 ISSN 2196-1735 (electronic)
Springer Series in Advanced Manufacturing
ISBN 978-3-030-77106-5 ISBN 978-3-030-77107-2 (eBook)
https://doi.org/10.1007/978-3-030-77107-2

This Springer imprint is published by the registered company Springer Nature Switzerland AG
The registered company address is: Gewerbestrasse 11, 6330 Cham, Switzerland

Preface

This book explores, in a systematic way, both conventional and unconventional material shaping processes with various modes of hybridization in relation to theory, modelling and industrial potential. In general, it summarizes the state-of-the-art hybrid manufacturing processes based on available literature sources and production reports. It is needed because the demand for high productivity and high accuracy is continuously increasing based on improvement and optimization strategies. In this challenge, the hybridization of manufacturing processes will play a crucial role and will be of a key importance in achieving high level of environmental and economical sustainability. This book contains 14 chapters written by specialists in conventional and unconventional machining processes with visible contributions in scientific and practical literature. It is structured into three parts: the first part (Chaps. 1–5) provides information on the physical fundamentals of the removal and non-removal processes in macro-, micro- and nanoscales, the second part (Chaps. 6–12) overviews the possible ways of hybridization and the effects on the enhancement of process performance, and the third part (Chaps. 13 and 14) summarizes production outputs related to process efficiency, surface finish and surface integrity, specifically with respect to difficult-to-machine materials. In particular, topics discussed consider the applications of different sources of hybridization including mechanical, thermal and chemical interactions or their combinations. All processes discussed are illustrated by appropriate schemes, photographs, diagrams and tables.

Chapter 1 provides general characteristics of material shaping processes including traditional and non-traditional machining processes which belong to the group of removal (subtractive) processes. In contrast, additive processes which utilize various material layer deposition techniques are characterized. The third group considers transformative processes, termed also non-removal processes, which are based on the intensive plastic deformation of the surficial layer resulting in the modification of its mechanical properties. A short description of hybrid processes generated by different combinations of the mentioned subtractive, additive and transformative processes is given.

Chapter 2 outlines classification systems of hybrid machining processes based on various criteria. In particular, the classification system is proposed by CIRP which distinguishes assisted and combined processes, and processes with controlled mechanisms. Different energy sources, i.e. mechanical, thermal, thermo-mechanical and chemical, are taken into account. The general rule is to classify conventional and unconventional processes integrally.

Chapter 3 overviews possible applications of hybrid machining processes in various industrial sectors, i.e. automotive, aerospace, precision, die and mould making, etc. They are also extended to hybrid micro- and nano-processes which are frequently used in fabrication of miniature parts (MEMS). In addition, the role of hybrid processes in sustainable manufacturing and Production 4.0 strategy, which is very important for manufacturing development, is discussed. Future vision of the development of hybrid manufacturing processes is proposed based on authorities in manufacturing theory and practice.

Chapter 4 provides physical fundamentals of conventional and unconventional machining processes which are crucial for understanding more complex mechanisms of hybrid processes and their synergy and interactions.

The following groups of conventional and unconventional processes are considered:

- Cutting processes
- Abrasive processes
- Electrical discharge machining (EDM)
- Electrochemical machining (ECM)
- Water jet machining (WJM) and abrasive water jet machining (AWJM)
- Laser beam machining (LBM)
- Ion beam machining (IBM) and electron beam machining (EBM)

As mentioned in Chap. 2, different types of energies and mechanisms of material removal, i.e. mechanical, thermal and chemical mechanisms, were considered.

Chapter 5 is devoted to modelling of machining processes which are used as basic components of hybrid machining processes. According to the strategy used in Chap. 4, the first group deals with modelling of cutting, grinding and finish abrasive processes and the second group concerns modelling of uconventional processes (EDM, ECM, LBM, WJM). Basic modelling techniques (analytical, numerical and AI-based) are shortly characterized.

Chapter 6 is devoted to vibration-assisted machining processes in which mechanical energy of vibrations (typically at the ultrasonic frequency (US)) assists the main machining process. In this chapter, the following processes are described in terms of process conditions, resonant devices and tooling.

- Vibration-assisted cutting
- Vibration-assisted grinding and polishing
- Vibration-assisted electrical discharge machining (EDM)
- Vibration-assisted electrochemical machining (ECM)

Chapter 7 is devoted to media-assisted machining processes in which mechanical energy of different media such as fluids, gases and cryogenic gases is used to improve process performance and achieve higher productivity. Three types of media assistance for the most popular cutting processes (turning and milling) are considered, namely:

- Fluid media-assisted machining
- Gaseous media-assisted machining
- Cryogenic machining

Chapter 8 is devoted to applications of magnetic and electric fields which predominantly assist unconventional (EDM and ECM) and finish abrasive processes in order to improve circulation of dielectric fluid or electrolyte and, as a result, enhance both the productivity and surface finish. The following cases are outlined:

- Magnetic field-assisted finishing processes
- Magnetic field-assisted electrical discharge processes
- Magnetic field-assisted electrochemical processes
- Electric field-assisted processes

Chapter 9 overviews the third basic method of hybridization which utilizes additional sources of thermal energy, i.e. laser beam, plasma beam or torch, electric torch and induction. It should be noted that laser is the most popular and the most effective source of thermal energy which assists practically all manufacturing processes in macro-, micro- and nanoscale. Thermally assisted machining processes are divided into the following groups:

- Laser-assisted processes
- Induction-assisted processes
- Plasma-assisted processes

Chapter 10 is devoted to the so-called mixed hybrid processes which combine two processes with the same or different material removal mechanisms. In practice, the following combinations of componential processes are used and included into this book:

- Combination of subtractive conventional processes (for instance, turning and broaching)
- Combination of subtractive conventional and non-conventional processes (for instance, grinding and EDM)
- Combination of subtractive unconventional processes (for instance, EDM and ECM)

Their importance results in improved process performance and producing defect-free surfaces on parts made of difficult-to-machine materials.

Chapter 11 considers special group of hybrid processes in which controlled mechanism of a basic process, e.g. grinding, leads to an improvement of mechanical properties of the subsurface layer. They are used in mass production of, for instance, geared shafts and eliminate classical heat treatment processes. Their application leads to substantial energy saving and producing parts with high functional properties, e.g. fatigue strength. The following combinations are described:

- Grind hardening and strengthening
- Combination of rolling and hardening
- Combination of subtractive and transformative processes

Chapter 12 is devoted to a relatively new group of hybrid processes which uses a sequence of additive and subtractive processes. Its wider application in advanced manufacturing sectors, such as aerospace and automotive, results from the fact that complex 3D parts are fabricated fully or partly by means of various additive techniques. Subtractive process is used as a finishing operation. This type of process hybridization is dynamically developed due to economic and environmental reasons. The following cases are described:

- Applications of different additive technologies and subtractive processes
- Repair and renovation technologies
- Special hybrid manufacturing platforms

Chapter 13 considers economics and optimization strategies of hybrid processes although these problems are not solved satisfactorily. The authors rely on some selected case studies from different manufacturing sectors. However, the general benefits are quite evident.

Chapter 14 concerns the influence of process hybridization on surface integrity including alterations of machined surface and subsurface layer. This problem was discussed systematically throughout previous chapters and summarized in relation to relevant hybridization methods. Problems such as surface topography, residual stresses and material alterations in the subsurface zone are considered. Similar to Chap. 13, the general problem considered is still open.

This new book addresses the present state and prospect of hybrid machining and other material shaping technologies. It can inspire current and future research activities in academic laboratories and R&D centres. This book is a comprehensive source of engineering-oriented information on hybrid manufacturing processes which are applied on various fabrication scales and in different manufacturing sectors. Moreover, it will be a valuable companion to many advanced books on manufacturing processes and systems and researchers will find it a good source of reference.

Opole, Poland — Wit Grzesik
Kraków/Nowy Sącz, Poland — Adam Ruszaj

Contents

Symbols and Abbreviations

Latin Symbols

a_p	Depth of cut (in-feed)
A	Vibration amplitude, material absorptivity, constant in J-C model
b	Width of cut
B	Magnetic induction
c	Specific heat
C	Specific volumetric heat
C_H	Volumetric hydrogen concentration
$\mathrm{C_u}$	Total cost per piece
d	Wire diameter in WEDM
e_c	Specific cutting energy
E	Cutting energy, Young modulus, sum of electric potential drops in near-electrode layers
E_{chem}	Chemical energy
E_{mech}	Mechanical energy
E_{therm}	Thermal energy
f	Feed rate
F_c	Cutting force
F_f	Feed force
F_p	Passive force
h	Uncut chip thickness, coefficient of natural heat convection
h_{ch}	Chip thickness
h_{lim}	Limited uncut chip thickness
h_{ig}	Sublimation heat
h_{min}	Minimum thickness of removed layer
I_p	Electric current amplitude
k	Heat conductivity of workpiece material
k_h	Chip compression ratio
k_c	Specific cutting pressure

k_v	Volumetric electrochemical equivalent of workpiece material
m	Mass
$\dot{m}_s$	Flow mass intensity in AWJM
n	Spindle rotational speed
p	Liquid pressure
p_c	Specific cutting power
p_n	Saturated electrolyte vapour pressure in assumed flow conditions
P_c	Cutting power
P_r	Production rate (efficiency)
P_{sh}	Shear plane
q_{ch}, q_m, q_t	Heat flux transferred to chip, material and tool
$q_{ch}, q_{cool}, q_m, q_w$	Heat flux transferred to chips, coolant, workpiece material and grinding wheel
Q	Cutting heat
Q_v	Volumetric material removal rate
r_n (r_β)	Cutting edge radius
t_u	Total time per piece
t_m	Machining time
T	Temperature
T_c	Cutting temperature
T_{melt}	Melting temperature, temperatura topnienia
u	Laser beam scanning speed
v_c	Cutting speed
v_f	Speed of electrode–tool displacement in ECM
v_w	Water speed in WJM
VB_B	Width of flank wear
V_v	Volumetric material removal rate

Greek Symbols

α	Linear coefficient of material extensibility, angle of outline, diffusion factor (coefficient)
α_o	Orthogonal clearance angle
β_o	Orthogonal wedge angle
ε	Linear deformation
$\dot{\varepsilon}$	Strain rate
γ	Shear strain
γ_o	Orthogonal rake angle
ϕ	Shear angle
λ	Thermal conductivity
ν	Poisson coefficient
ρ	Material density, radius of abrasive grain
ρ_w	Water density

σ	Residual stress
σ_{eq}	Equivalent flow rate in J-C model

Abbreviations

1D	One-dimensional (linear) system
2D	Two-dimensional system
3D	Three-dimensional system, printing
AECS	Abrasive-assisted electrochemical smoothing
ADL	Aerosol dry lubrication
A(A)EDM	Abrasive-assisted electrodischarge machining
AAEM	Abrasive-assisted electrochemical machining
AFM	Abrasive flow machining
AI	Artificial intelligence
AJM	Abrasive jet machining
AM	Additive machining/manufacturing
ANNs	Artificial neural networks
ASJM	Abrasive slurry jet machining
AWEDM	Abrasive wire electrodischarge machining
AWJM	Abrasive water jet machining
CAD	Computer-aided design
CAM	Computer-aided manufacturing
CBN	Cubic boron nitride
CF	Cutting fluid
CDR	Cryogenic deep rolling
CHM	Chemical machining
CIRP	International Academy of Production Engineering
CNC	Computer numerical control
CT	Conventional turning
D	Drilling
DC	Direct current
DED	Direct energy material deposition
DMP	Direct melted printing
EBM	Electron beam melting
ECG	Electrochemical grinding
ECJM	Electrochemical jet machining
ECM	Electrochemical machining
ECDM	Electrochemical–electrodischarge machining
ECSM	Electrochemical spark machining
FAM	Fluid-assisted machining
EDAG	Electrodischarge abrasive grinding
EDM	Electrodischarge machining
EM	Emulsion

FDM	Fused deposition modelling
FEM	Finite element method
FFBP	Feedforward backpropagation neural network
FIBM	Focused ion beam machining
FL	Fuzzy logic
G	Grinding
GH	Grind hardening
HAZ	Heat-affected zone
HB	Brinell hardness
HM	Martens hardness
HM	Hard machining
HP	Hybrid production
HPC	High-pressure cutting
HPJAM	High-pressure jet-assisted cutting
HPWJAM	High-pressure water jet-assisted machining
HRC	Rockwell (C) hardness
HSC	High-speed cutting
HSM	High-speed machining
IAM	Induction-assisted machining
IEG	Inter-electrode gap
IBM	Ion beam machining
IR	Infrared radiation
LAG	Laser-assisted grinding
LAJECM	Laser-assisted jet electrochemical machining
LAM	Laser-assisted machining
LAMG	Laser-assisted micro-grinding
LAMM	Laser-assisted micro-milling
LBDW	Laser beam deposition welding
LBM	Laser beam machining
LCA	Life cycle assessment
LAECM	Laser-assisted electrochemical machining
LAEDM	Laser-assisted electrodischarge machining
LAG	Laser-assisted grinding
LAM	Laser-assisted machining
LAMILL	Laser-assisted milling
LAT	Laser-assisted turning
LCO_2	Liquid carbon dioxide
LDT	Laser deposition technology
LMD	Laser metal deposition
LN_2	Liquid nitrogen
LRT	Laser repair technology
LS	Laser scanning
MFAECM	Magnetic field-assisted electrochemical machining
MFAEDM	Magnetic field-assisted electrodischarge machining
M(MILL)	Milling

MAM	Media-assisted machining
MDM	Multi-nozzle deposition manufacturing
MEMS	Microelectromechanical system
MF	Magnetic field
MMCs	Metal matrix composites
MQC	Minimum quantity cooling
MQL	Minimum quantity lubrication
MQCL	Minimum quantity cooling–lubrication
MPECM	Mixed powder electrochemical machining
MRR	Material removal rate
NEMS	Nanoelectromechanical system
PAM	Plasma-assisted machining
PECM	Pulsed electrochemical machining
PFB	Powder fusion bed
PBM	Plasma beam machining
PCBN	Polycrystalline cubic boron nitride
PCD	Polycrystalline diamond
PLP	Pulsed laser plasma
PMEDM	Powder-mixed electrodischarge machining
RS	Residual stress
RUM	Rotary ultrasonic machining
RUAG	Rotary ultrasonic-assisted grinding
SACE	Spark-assisted chemical engraving
SEM	Scanning electron microscopy
SEDCM	Simultaneous micro-EDM and micro-ECM milling
SGS	Surface geometrical structure
SI	Surface integrity
SLA	Stereolithography
SLS	Selective laser sintering
SLM	Selective laser melting
SM	Subtractive machining, sustainable manufacturing
SHPB	Split Hopkinson pressure bar
T	Turning, temperature in K
TAM	Thermally assisted machining
TSL	Technological surface layer
TWS	Tool–workpiece separation in UAMILL
UAD	Ultrasonic-assisted drilling
UAM	Ultrasonic-assisted machining
UAMILL	Ultrasonic-assisted milling
UAT	Ultrasonic-assisted turning
US	Ultrasonic vibration
USM	Ultrasonic machining
USAM	Ultrasonic-assisted machining
USAD	Ultrasonic-assisted drilling
USAT	Ultrasonic-assisted turning

USAECM	Ultrasonic-assisted ECM
USAEDM	Ultrasonic-assisted EDM
USAG	Ultrasonic-assisted grinding
UV	Ultraviolet radiation
VAM	Vibration-assisted machining
VAMILL	Vibration-assisted milling
WJM	Water jet machining
WSACE	Wire spark-assisted chemical engraving

Chapter 1
General Characteristics of Material Shaping Processes

1.1 Classification of Material Removal Processes

Material removal process (machining) is one of the five basic groups of all manufacturing technologies presented in Fig. 1.1. They include the processes of joining, separating and removal (subtractive) shaping, non-removal shaping (processing, transforming) and additive shaping [1, 2]. This classification was proposed in connection with the intensive development of hybrid manufacturing processes, which has been visible in recent years. In the literature on material removal (subtractive) processes, the classification into conventional machining and erosive (unconventional) is usually used (Fig. 1.2) [1]. As shown in Fig. 1.1, but also from the actual state of the manufacturing industry, the additive shaping processes that limit, and in many cases eliminate, the conventional removal and non-removal processes are of significant importance.

The basic manufacturing techniques currently used for the shaping of machine and technological components are *non-removal* shaping, which includes casting and forming, as well as *subtractive* machining, including the processes classified in Table 1.1. The mass reduction of a semi-finished product can generally be carried out using mechanical, thermal and chemical energy, although the interaction of two or more types of energy, as in the case of *hybrid* machining, becomes increasingly popular [3].

It should be emphasized that the *hybridization of* manufacturing processes, including the material removal processes, currently plays an important role in achieving a high level of manufacturing innovation through the intended development of the applied manufacturing processes, as it creates large, additional opportunities for their improvement and optimization.

Table 1.1 lists the material removal processes used on an industrial scale that can be further hybridized (see Table 1.2 in this Chapter) according to different concepts. These processes—conventional and unconventional, using different energy sources (mechanical, thermal, chemical and electrochemical)—can be integrated into more efficient and effective hybrid processes using two (or more) energy sources or different mechanisms of removing excess material.

W. Grzesik and A. Ruszaj, *Hybrid Manufacturing Processes*, Springer Series in Advanced Manufacturing, https://doi.org/10.1007/978-3-030-77107-2_1

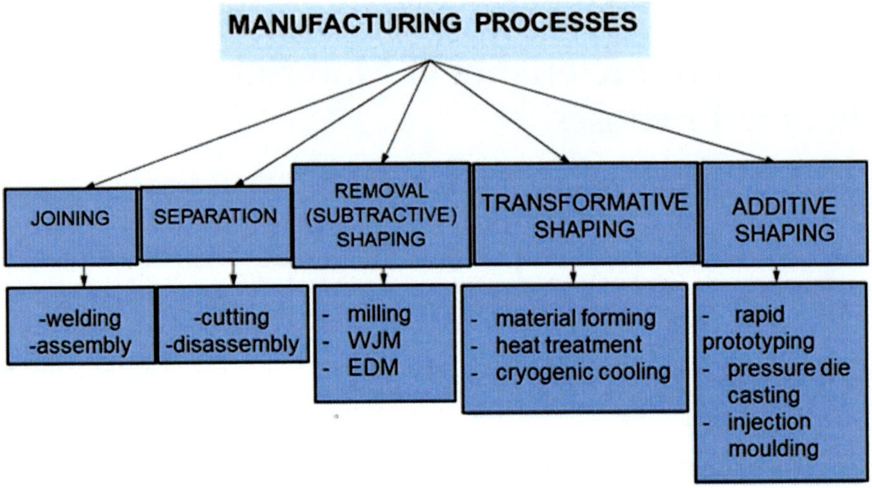

Fig. 1.1 Specification of groups of manufacturing processes [2]

Fig. 1.2 Classification of material removal/subtractive processes [1]

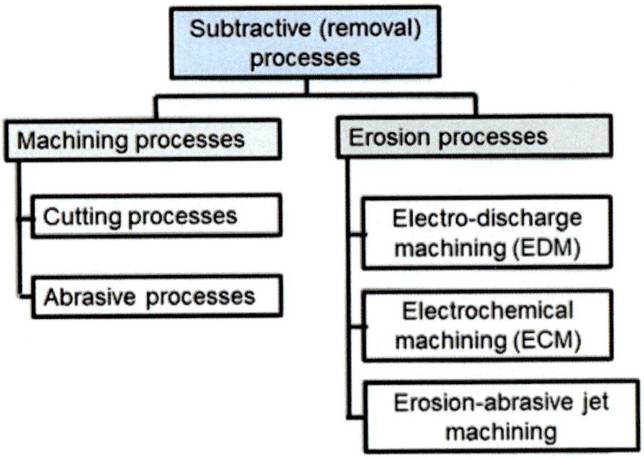

1.2 Classification of Additive Processes

Additive techniques, popularly known as *3D printing*, are usually divided according to the form of the material used (building material) and the way it is joined and fed [5]. In the case of the material criterion (Fig. 1.3), it is important whether the material used was solid or in powder form, or as liquid, i.e. in liquid or semi-liquid form [6]. As can be seen in Fig. 1.3, depending on the form of the input material, different joining techniques are used, including sintering, melting and bonding. The classification based on three methods of applying material layers, i.e. by bonding the layers, forming on a platform and depositing is shown in Fig. 1.4.

If the method of material supplying is considered to the basic criterion (Fig. 1.5),

Table 1.1 Classification of advanced machining processes (AMP) due to the energy used to remove material, giving the source of energy, tool and medium of energy transfer, and material removal mechanism [4]. (www.link.springer.com)

Type of energy	AMP	Energy source	Tool	Energy transfer medium	Material removal mechanism
Mechanical	USM	Ultrasonic vibration	Sonotrode	Abrasive slurry	Erosion and/or abrasion
	AJM	Pneumatic pressure	Abrasive jet	Air	
	WJM	Hydraulic pressure	Water jet	Air	
	AWJM	Hydraulic pressure	Water-abrasive planer	Air	
	IJM	Hydraulic pressure	Ice particles jet	Air	
	AFM	Hydraulic pressure	Abrasive suspension	Suspension	
Chemical	CHM	Corrosive agent	Mask	Etchant	Chemical dissolution
Electrochemical	ECM	High current (I)	Electrode	Electrolyte	Anode dissolution by ion displacement
Thermal	EDM	High voltage (A)	Electrode	Dielectric	Melting and vaporization
	EBM	Ionized material	Electron beam	Vacuum	
	IBM	Ionized material	Ion beam	Atmosphere	
	LBM	Amplified light	Laser beam	Air/ atmosphere of special composition	
	PAM	Ionized material	Plasma jet	Plasma	

Symbols: *AFM*–abrasive flow machining, *AJM*–abrasive jet machining, *AWJM*–abrasive water jet machining, *CHM*–chemical machining, *ECM*–electrochemical machining, *EDM*–electrodischarge machining, *IBM*–ion beam machining, *IJM*–ice jet machining, *LBM*–laser beam machining, *PBM*–plasma beam machining, *USM*–ultrasonic machining, *WJM*–water jet machining

one can distinguish between shaping with the use of a platform filled with powder, called according to ASTM International classification the *PDF—powder bed fusion*, and *direct deposition*, which is used in sintering, melting and depositing processes. It should be added that the American Society for Testing and Materials (ASTM) group "ASTM F42—Additive Manufacturing", classifies additive processes in 7

Table 1.2 Classification of AM methods according to ASTM—their commercial variations [6–8]

AM process groups	Typical commercial names
Vat photopolymerization	Stereolithography, digital light processing, solid ground curing, projection stereolithography
Powder bad fusion	Electron beam melting, electron beam additive manufacturing, selective laser sintering, selective heat sintering, direct metal laser sintering, selective laser melting, laser beam melting
Direct energy deposition	Laser metal deposition, direct metal deposition, direct laser deposition, laser engineered net shaping, electron-beam freeform fabrication, welded-based additive manufacturing
Binder jetting	Powder bed and inkjet head, plaster-based 3D printing
Material extrusion	Fused deposition modelling, fused filament fabrication
Material jetting	Multi-jet modelling
Sheet lamination	Laminated object manufacturing, ultrasonic consolidation

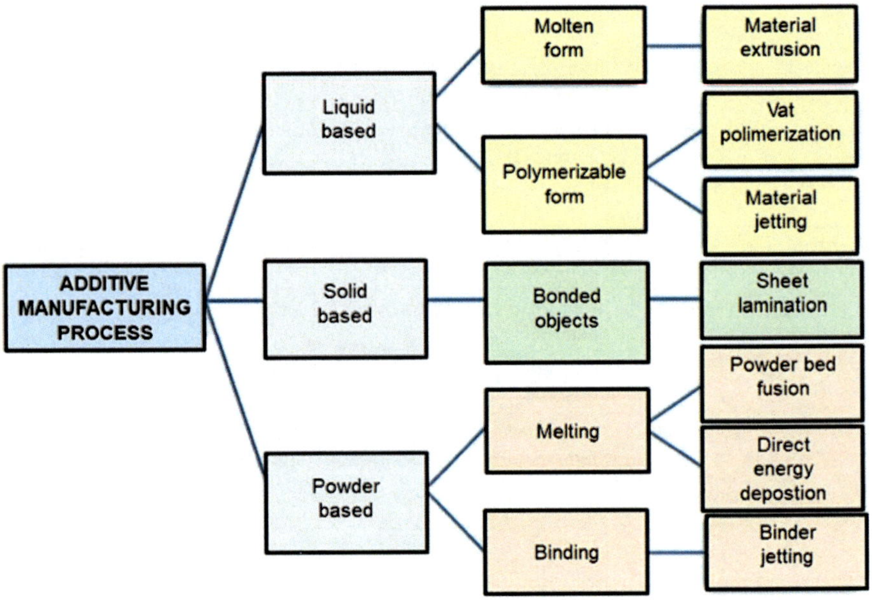

Fig. 1.3 Classification of additive processes in terms of initial material type and joining method [6] [ww.omnexus.specialchem.com]

groups (Standard Terminology for Additive Manufacturing Technologies, 2012). Their acronyms and practical names are listed in Table 1.2.

The advantage of using evenly distributed powder on the working platform is that it is not necessary to produce the supports beforehand. The metal powder bed technology uses a high power ytterbic fiber laser to melt fine metal powders to form useful three-dimensional parts. It is often referred to as the layered melting method,

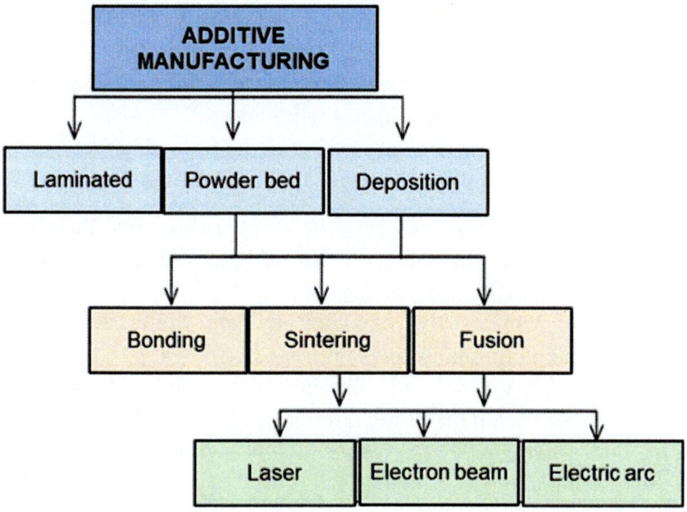

Fig. 1.4 Classification of additive processes in terms of material deposition method (www.raiith.iith.ac.in/1019/1/2013-75)

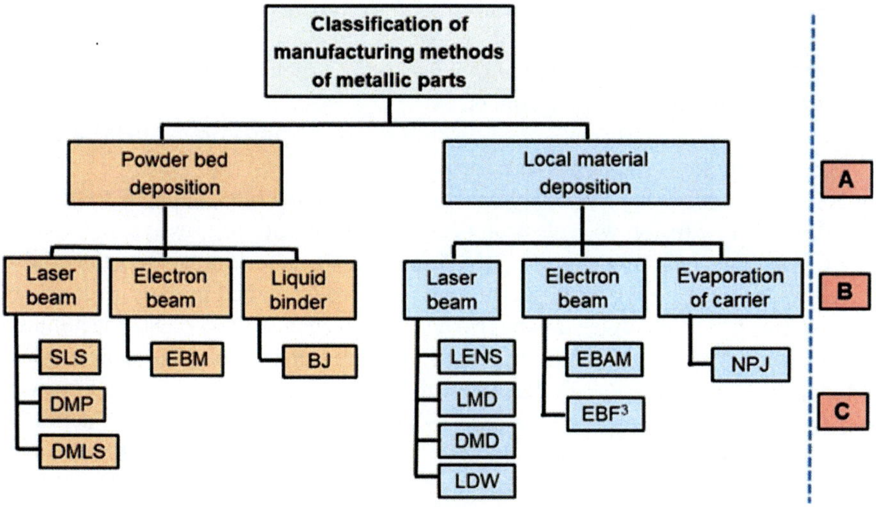

Fig. 1.5 Classification of additive processes for metallic parts in terms of: **a** material supplying, **b** joining, **c** typical acronyms (www.mechanik.media.pl)

the technology of additive metal manufacturing, metal 3D printing, laser melting or additive metal manufacturing using metal. The manufacturing process uses 2D sections of a 3D CAD model. Each cross-section of the CAD model corresponds to a thin layer of fine metal powder placed on the working surface; then selected areas of the powder are precisely melted by laser. This process is repeated layer by layer until the production of the object is completed. Additive manufacturing systems can use many metallic materials, e.g. titanium alloy Ti6Al4V, cobalt alloys, chromium, stainless steel, nickel alloys—Inconel 625, Inconel 718 and aluminum alloy AlSi10Mg or some mixtures of metallic powders.

According to the adopted classification criteria, the following methods of additive manufacturing of wider practical importance can be distinguished [5, 9, 10]:

- SLS, i.e. *selective laser sintering* of powders of various metal materials. Powder particles are combined as a consequence of their melting, which results in increased porosity of the produced elements.
- SLM, i.e. *selective laser melting* of metal powders. Material is fully melted in selected areas, which eliminates the occurrence of pores.
- DMP, i.e. *direct metal printing*, which means direct molten metal printing in the areas where powder stream meets the laser beam. It is a trademark of 3D Systems.
- DMLS, i.e. *direct metal laser sintering*, which uses direct metal powder sintering by laser. It is similar to the SLS method but is used to produce parts from a mixture of different powders.
- BJ, i.e. *binder jetting*, i.e. spraying powder particles with binder.
- NPJ, i.e. *nanoparticle jetting*, which stands for the jet spraying of a mixture of nanoparticles of material and liquid carrier that settles in the form of drops. The thickness of a single layer is reduced to 2 μm. This technology is patented by Xjet.
- EBM, i.e. *electron beam melting*, consisting in melting the material with an electron beam in a vacuum chamber.
- EBAM, i.e. *electron beam additive manufacturing*, similar to the EBM method, but the material is fed in the wire form.
- EBF^3 (*electron-beam freeform fabrication*), which is used by NASA to produce 3D models by fusion with electron beams in the absence of gravity.
- LC, i.e. *laser cladding*, based on the laser deposition technique. The material is fed in the form of powder/wire and then sintered or melted by laser. It is one of the basic AM methods used in hybrid techniques such as (AM+CNC machining), including also part repairs. It is actually a family of various techniques such as LMD, LENS, DMD and LDMD. They will be described in more detail in Chap. 12.
- LDW, i.e. *laser deposition welding*, which is laser deposition of material by welding. It is a technique used on a wider scale to repair or add additional elements to the part shape by DMG Mori company.
- LMD, i.e. *laser metal deposition*, which consists in laser deposition of material as a result of powder (LMD-p) or wire (LMD-w) laser deposition.
- LENS, i.e. *laser engineering net shaping*, which is a variation of the LC method used for final laser cladding of parts. This method was developed by Sandia Corporation.
- DMD, i.e. *direct metal deposition*, which means the direct cladding/deposition of material. It is also one of the varieties of the LC methods.

Undoubtedly, the most commonly used in industry, and the simplest and cheapest method is FDM (*Fused Deposition Modeling*), which involves depositing successive layers of molten thermoplastic. The method allows to make models with different properties due to the use of different materials, including ABS, PLA, PC, Nylon, TPU and more exotic composites based on glass fiber, carbon fiber or admixture of wood dust. It is particularly suitable for the rapid production of individual parts, which is the reason why designers often use it for project evaluation. However, the manufactured elements are characterized by low strength, due to the unfavorable, layered internal structure.

Another group includes methods using photosensitive resins, including SLA (*Stereolithography*), DLP (*Direct Light Processing*), CDLP (*Continuous DLP*). These consist of hardening resins with different types of light and are used in industry and more and more often also in amateur 3D printing. Thanks to high resolution they allow producing small and accurate models with high surface quality as well as a large variety of obtainable properties.

Additive technologies also include powder sintering, which is a group of following methods: SLS (*Selective Laser Sintering*), SLM/DMLS (*Selective Laser Melting/Direct Metal Laser Sintering*) and EBM (*Electron Beam Melting*). They consist in sintering successive layers of material powder with a concentrated energy beam (of electrons or laser). These processes use durable polymeric materials (e.g. polyamides, TPU), composite materials (e.g. PA+GF) and metallic materials (various mixtures of metal powders, e.g. stainless steel, titanium, aluminum). Their important advantage is that they do not require support structures for prints, which allows obtaining more complex shapes. These methods work well when it is necessary to produce many parts in the shortest possible time, because the entire workspace is occupied by models and which are all produced in one process. Other advantages include productivity (about 10 times higher than in FDM technology) and high strength. The parts produced in this way have a practically homogeneous internal structure.

The popular methods are also *Material Jetting*, among others, industrially used *PolyJet* and *Multi Jet Fusion*. The *PolyJet* method consists in applying successive layers of material and then hardening them with UV light. This takes place during one pass of the head, so the printing process is fast. Advanced machines also allow printing with dyeing in different colors and from different fillers/filaments. The surfaces obtained are of high quality, therefore these technologies are used to produce realistic prototypes or high quality short product batches. However, there are limitations due to high cost of machines and materials.

In practice, industry uses processes based on direct energy supply, including LENS (*Laser Engineering Net Shape*) and EBAM (*Electron Beam Additive Manufacturing*). Their special feature is that the material is supplied by a special nozzle and hardened by laser or electron beam directly on the surface it falls on. Important advantages include the ability to print on surfaces of any shape and a wide range of materials.

References

1. Grzesik W (2017) Advanced machining processes of metallic materials. Elsevier Amsterdam
2. Zhu Z, Dhokia VG, Nassehi A, Newman ST (2013) A review of hybrid manufacturing processes. Int J Comp Integr Manuf 26(7):596–615
3. Lauwers B, Klocke F, Klink A, Tekkaya AE, Neugebauer R, Mc Intosh D (2014) Hybrid processes in manufacturing. CIRP Ann Manuf Technol 63(2):561–583
4. Gupta K, Jain NK, Laubscher RF (2016) Hybrid machining processes. Perspective on machining and finishing. Heidelberg Springer
5. Gibson I, Rosen DW, Stucker B (2010) Additive manufacturing technologies
6. 3D Printing/Additive manufacturing using polymers—complete guide. https://omnexus.specialchem.com
7. Stavropoulos P, Foteinopoulos P (2018) Modelling of additive manufacturing processes: a review and classification. Manuf Rev 5(2):1–26
8. Zhang Y, Linmin W, Xingye G, Kane S, Deng Y, Jung Y-G, Lee J-H, Zhang J (2018) Additive manufacturing of metallic materials: a review. J Mat Eng Perform 27:1–13
9. Schmidt M, Merklein M, Bourell D, Dimitrov D, Hausotte T, Wegener K, Overmeyer L, Vollertsen F, Levy GN (2017) Laser based additive manufacturing in industry and academia. CIRP Ann Manuf Technol 66(2):561–583
10. Kozak J, (2018) Mathematical modelling of advanced manufacturing processes. Science library of the institute of aviation no. 56, Warsaw

Chapter 2
Classification of Hybrid Machining Processes Generating by Different Rules

2.1 Classification Criteria and Systems

The evolution of conventional and unconventional machining processes in the period after the Second World War is inextricably linked to the combination of individual processes and the use of different active energy sources or the implementation of several machining methods or even several subsequent stages of the technological process, in a single production unit in order to achieve the synergy effect. This means that a hybrid process, a hybrid machine tool or a manufacturing platform contributes to the resultant effect exceeding the sum of the effects of the componential processes carried out separately.

Basically, the material removal componential processes listed in Table 1.1 (Sect. 1.1) can be integrated into more efficient and effective hybrid processes as shown in Table 2.1 using two (or more) energy sources or various material removal mechanisms. It should be noted that there is a new terminology for the hybridization of manufacturing processes introduced into the literature in the last decade.

In a natural way, the definition of hybrid machining has evolved due to the development of techniques and ways of manufacturing, which are inaccurately called manufacturing technologies now. The first definition from the 1970s and their later versions considered the combination of two or more processes that can be used for material removal at the same time (e.g. a combination of abrasive removal with EDM or ECM), or the fact that one of them is only of an auxiliary nature, contributing to a beneficial change in process conditions, e.g. cutting with laser material heating or cryogenic cooling. Nowadays, hybridization includes various manufacturing techniques, i.e. removal, non-removal processes, (material transformation, change of microstructure and mechanical properties) and additive machining and their combinations [1–3]. In order to unify the terminology, it has been proposed to change their names to *subtractive machining*, *transformative machining* and *additive machining* [1, 2].

The definition of hybrid manufacturing processes currently recommended by CIRP, which includes manufacturing/machining processes, is as follows [3]:

W. Grzesik and A. Ruszaj, *Hybrid Manufacturing Processes*, Springer Series in Advanced Manufacturing, https://doi.org/10.1007/978-3-030-77107-2_2

> Hybrid manufacturing processes are based on simultaneous and controlled interaction of process mechanisms and/or energy sources/tools having a significant effect on the process.

The term *simultaneous and controlled interaction* means that processes/energy sources must interact, more or less, in the same zone of the hybrid process and at the same time. The reinforcement of the total effect of the process hybridization is represented by the principle ("$1 + 1 = 3$ effect"), which indicates the achievement of a more efficient machining process, e.g. by thermal softening of the workpiece material with a laser heating.

In general (Fig. 2.1), there are processes based on combining different energy sources or different tools (different shaping methods and operations) included in group *I* and processes using controlled mechanisms of different processes, which are realized in conventional componential processes (group *II*). Consequently, group I distinguishes *assisted processes* (subgroup *I.A*) and *mixed/combined processes* (subgroup *I.B*).

For conventional and unconventional subtractive machining, the most important thing is to create hybrid processes according to the *I.A.* principle (support of different cutting operations with vibration energy, thermal laser and liquid and gaseous media, also grinding, polishing, EDM, ECM) and *I.B* (e.g. combining grinding and EDM;

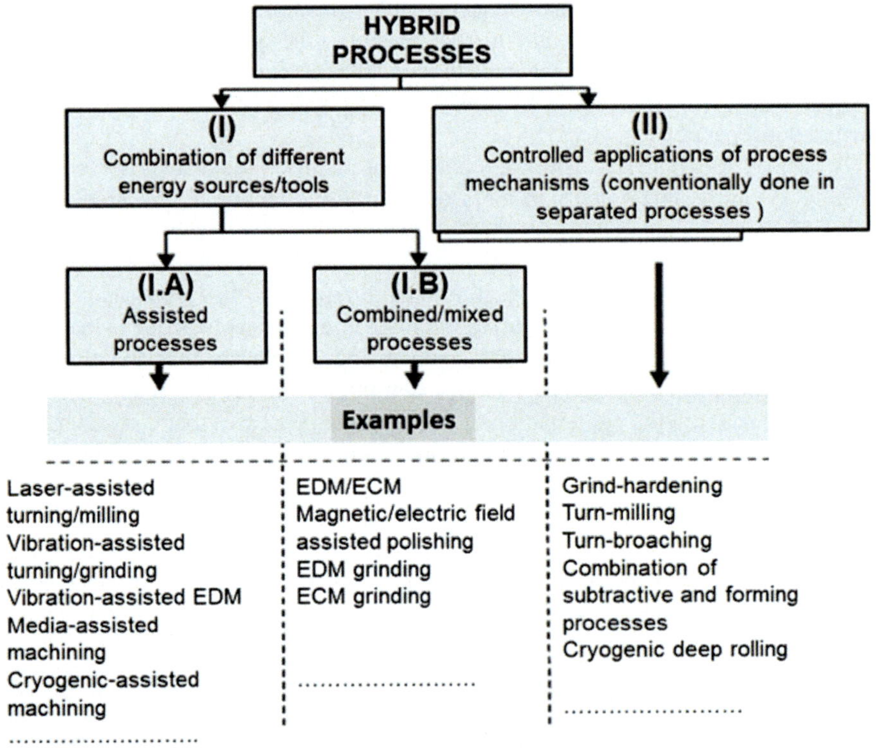

Fig. 2.1 Classification of hybrid manufacturing processes according to CIRP nomenclature [1, 3]

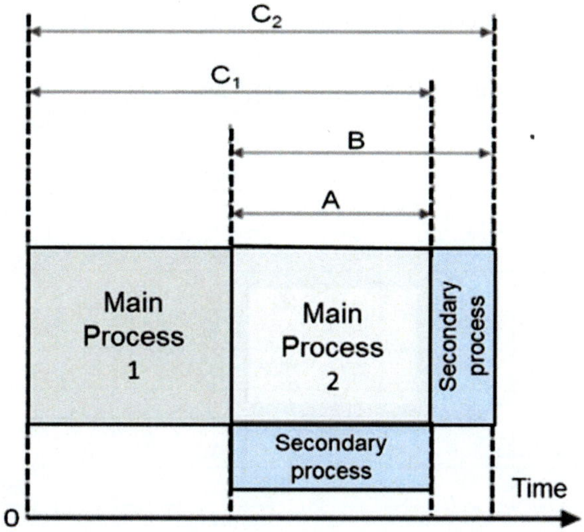

Fig. 2.2 Consideration of timing in hybrid machining processes [4]. Symbols: **A** concurrent application of main and secondary processes, **B** sequential application of main and secondary processes (M/S), $\mathbf{C_1}$ sequential application of two main processes (M/M), $\mathbf{C_2}$ sequential application of two main processes and one secondary process (M/M/S).

grinding and ECM; ECM and EDM). Other hybrid processes also concerning material forming are described in the work [2, 3].

In group II, examples can be given of combining the kinematic features of two cutting methods, e.g. *turn-milling* and *turn-broaching*, *grind-hardening*, *cryogenic deep rolling*, or combining machining with burnishing.

Figure 2.2 shows the time relations between the component processes for different varieties of hybrid processes [4] classified in Fig. 2.1. In varieties *I.A* and *II* the basic and supporting (secondary) process can be realized simultaneously (case *A*) or sequentially (case *B*), while the combination of two basic processes *1* and *2* is done sequentially (case C_1, which is well illustrated by the combination of initial cutting and finish burnishing). Case C_2, on the other hand, can be referred to the ECDG, in which, apart from EDM and ECM, there is also abrasive assistance. This means that a hybrid process can be assigned either to the process time or to the machining zone [3, 4].

2.2 Generation of Hybrid Conventional and Unconventional Processes

Table 2.1 summarizes the possible combinations of hybrid processes based on the classification of subtractive machining processes proposed in Table 1.1. It should be noted that in the group of classical machining methods based on the mechanical action

Table 2.1 Summary of possibilities of combining subtractive machining processes (hybridization of energy sources) [5]

		ED	LB	EB	PB	CH	EC	A	T	US	F
THERMAL	ED	EDM					ECDM	AEDG		EDUSM	
	LB		LBM			ELB	ECML		TLB		
	EB			EBM							
	PB				PPM				TPB		
	CH		ELB			CHM		EEM			CHP
	EC	ECAM	ECML				ECM	AECG		USECM	
MECHANICAL	A	AEDM				MCP	AECH	G		USG	
	T		TLB		TPB				T	UST	
	US	EDMUS	LBMUS				USMEC	GUS	TUS	USM	USP
	F					CHP	ECL			USP	JM

Possible combinations of subtractive machining methods. Symbols: A—abrasive tool, T—cutting tool, US—ultrasonic vibration, F—fluid

of a tool with a defined cutting edge geometry (T) on the workpiece material, process support applies to the *I.A.* principle. Therefore, the cutting process is assisted by laser (TLB), plasma (PLB) and ultrasonic vibration (UST). In turn, grinding (abrasive) support (A) concerns the principle of I.B and that is why it enables erosion (AEDM), chemical (MCP), electrochemical (AECH) and ultrasonic vibrations (USG) support. Unconventional processes (ED, CH, EC) can be combined, e.g. ECDM, ECAM and assisted by EDUSM, USECM, ECML.

Triangle diagrams (Fig. 2.3) may be used for better understanding the principles of hybridization of machining processes, showing the type of interaction of process

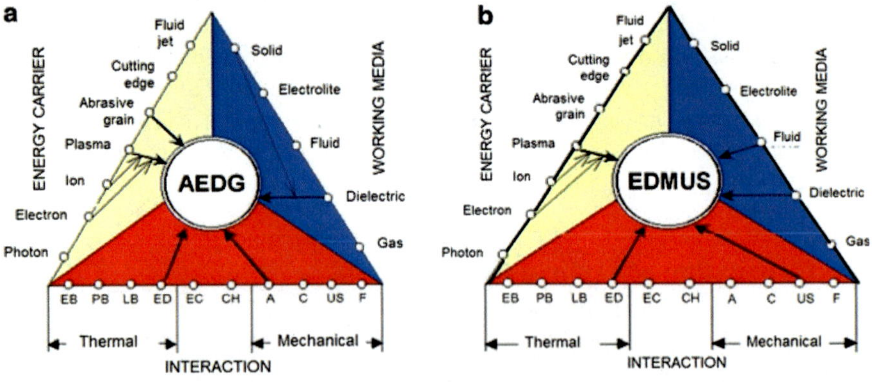

Fig. 2.3 Examples of energy structures of hybrid machining processes: **a** grinding and EDM (AEDG), **b** US vibration assisted EDM (EDMUS). (www.home.iitk.ac.in)

mechanisms depending on the working medium used and the energy carrier(s). Figure 2.3a and b give examples of abrasive machining assisted by electrodischarge (AEDG) and electrodischarge machining assisted by ultrasonic vibrations (EDMUS).

Figures 2.4 and 2.5 show the classification of hybrid subtractive machining processes into main (basic) and secondary [6], which in principle correspond to the

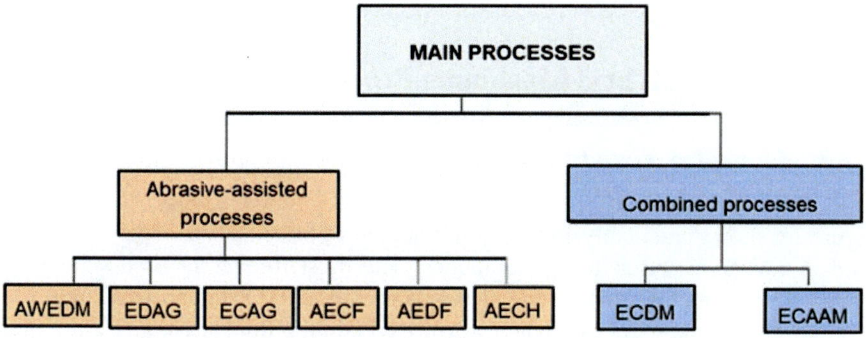

Fig. 2.4 Specification of main hybrid machining processes (HMP) [6]: AWEDM- abrasive wire EDM, EDAG–electrodischarge assisted grinding, ECAG–electrochemical assisted grinding, AECF–abrasive electrochemical finishing, AEDF–abrasive electrodischarge finishing, AECH–abrasive electrochemical honing, ECDM–electrochemical–electrodischarge machining, ECAM–electrochemical-arc (plasma) machining

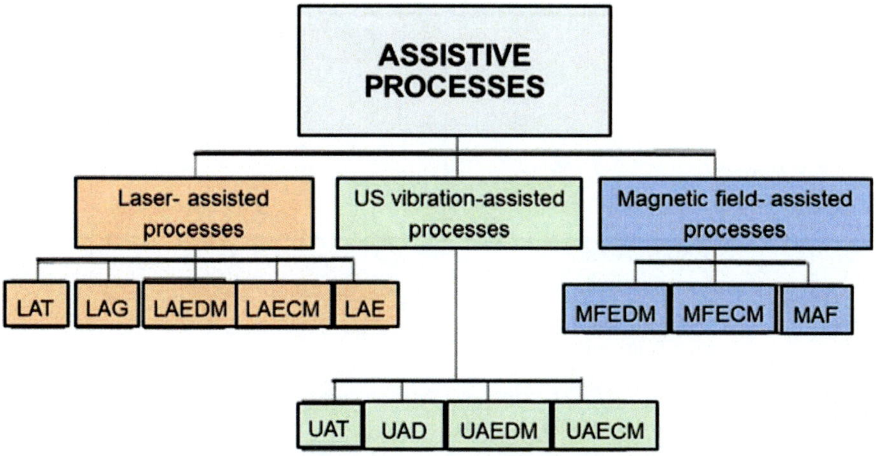

Fig. 2.5 Secondary hybrid machining processes (HMP) [6]: LAT–laser–assisted turning, LAG–laser—assisted grinding, LAEDM–laser-assisted electrodischarge machining, LAECM–laser—assisted electrochemical machining, MFEDM–magnetic field—assisted electrodischarge machining, MAF–magnetic field-abrasive finishing, UAT–US vibration—assisted turning, UAD–US vibration—assisted drilling, UAEDM- EDM US vibration—assisted EDM, UAECM–US vibration—assisted ECM

group of assisted processes (group I.A in Fig. 2.1) related to conventional and unconventional processes. The processes of abrasive material removal, e.g. [7, 8] assisted by chemical, electrochemical or electro-erosion effects have recently been limited to special alloys and composites used e.g. in the aircraft and aerospace industry (in the production of aeroengine components) due to harmful impact the working fluids on the working environment and the natural environment.

2.3 Survey of Hybrid Machining Processes

2.3.1 Hybrid Assisted Processes

Figure 2.6 shows practically applied combinations of basic and secondary processes used in various manufacturing techniques. The most frequently used supporting energy sources include: vibrations of frequency 0.1 ÷ 80 kHz and amplitude 1 ÷ 200 μm, liquid and gaseous media (pressurized cooling fluid (CF), liquid nitrogen (LN2), cooled CO_2) and laser [3, 9]. Therefore, the three most common groups of assisted processes (group I.A in Fig. 1.6) are *vibration/US-assisted machining*, *thermally-assisted machining* and *media-assisted machining*. The current applications of vibration-assisted processes in mass production and the state of research advancement are presented in Fig. 2.7.

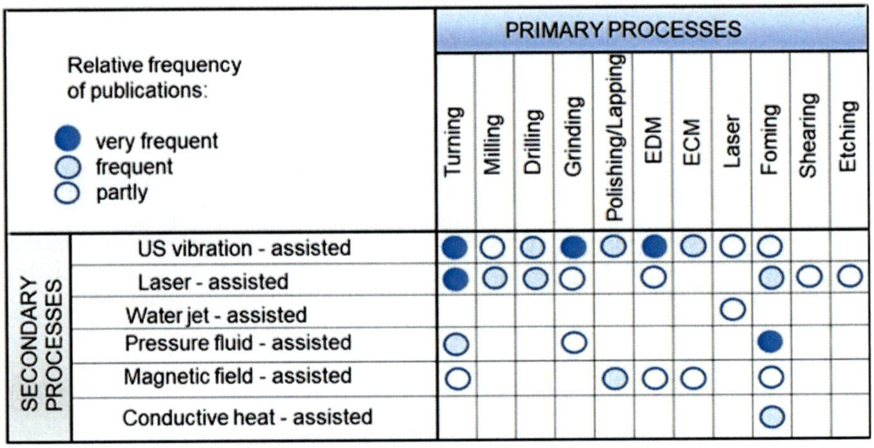

Fig. 2.6 Combinations of assisted manufacturing processes [3]

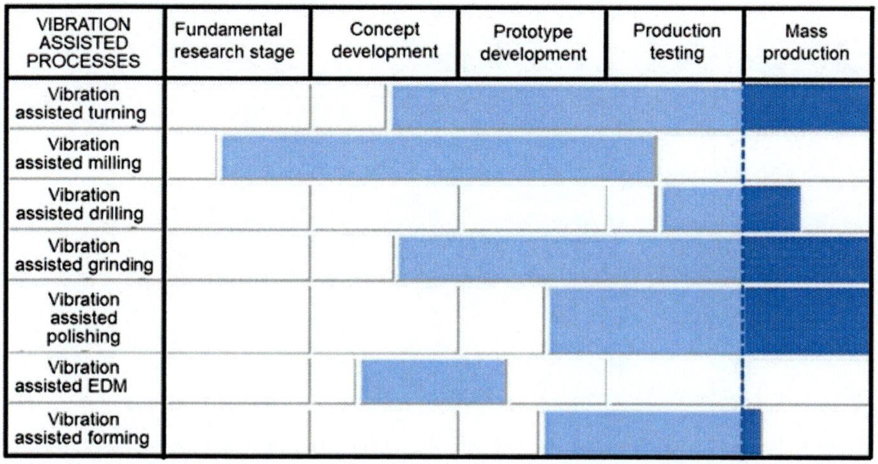

Fig. 2.7 State of the progress in investigations of hybrid vibration-assisted manufacturing processes [3]

2.3.2 Hybrid Combined Processes

As mentioned earlier, the combination of two or more subtractive machining processes is, according to the definition, determined by their simultaneous effect to a greater or lesser extent on the material removal mechanism (group *I.B* in Fig. 2.1). So far, the processes combining the abrasive grinding mechanism and the thermal effect of electrical discharges have found the widest application, i.e. AEDG (EDG) or electrochemical dissolution, i.e. AECG (ECG), and the combination of EDM and ECM, i.e. ECDM. In the last case, a third abrasive mechanism may occur, i.e. in a hybrid ECDG process. In the case of electrochemical abrasive machining, not only grinding (ECG) is carried out, but also honing (ECH) and superfinishing (ECS). The principles of combining the constituent processes into hybrid processes are shown in Fig. 2.8.

The review and analysis of the current state of hybrid machining allow drawing the following conclusions:

- *hybridization* of manufacturing processes, including subtractive machining, applies not only to typical cutting processes, but also to micro and nanoscale machining processes.
- hybrid machining processes produce, due to the synergy effect, the resultant effect exceeding the sum of the effects of separate componential processes. Therefore, additional possibilities for process optimization appear.
- in practice, hybrid processes assisted in the form of an additional energy source, combining different energy sources and/or tools and controlling different mechanisms of componential processes (subtractive, material forming, heat treatment, additive machining) can be used.

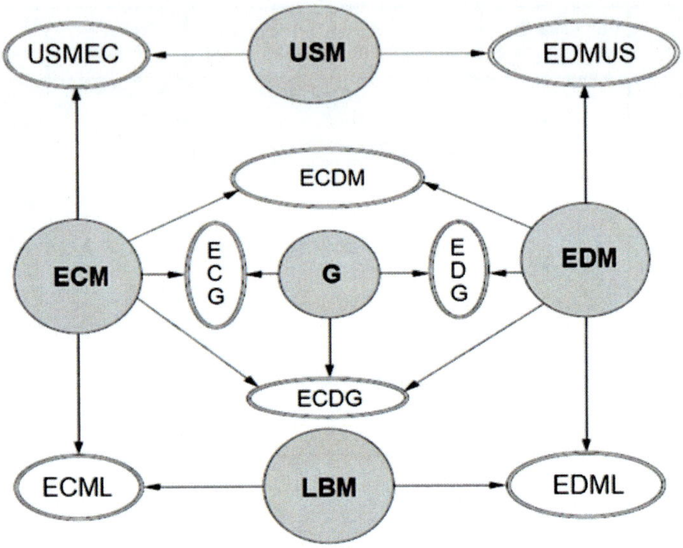

Fig. 2.8 Rules for combination of hybrid machining processes [5]

- in the group of conventional cutting methods (turning, drilling, milling), the most important is the support of the US vibration energy, laser and such technological media as high-pressure fluid and cryogenically cooled media. i.e. liquid nitrogen.
- in the group of conventional methods of abrasive machining (grinding, honing, polishing, lapping), the most important is to amplify the abrasive effect by means of electro-discharge and electrochemical action and magnetic field forces. However, the development of this group of hybrid processes is limited by ecological restrictions.
- for ecological reasons and requirements concerning the surface functionality of machine elements, hybrid processes are developed in which the mechanisms of the componential processes are controlled, e.g. controlled heat flow in hardening grinding, intensive surface layer deformation combined with cryogenic cooling and material phase transformation.
- there is a rapid development of hybrid processes and devices/machines, which combine additive shaping and CNC machining (to be included in Chap. 12).

References

1. Grzesik W (2017) Advanced machining processes of metallic materials. Elsevier Amsterdam
2. Zhu Z, Dhokia VG, Nassehi A, Newman ST (2013) A review of hybrid manufacturing processes. Int J Comp Integr Manuf 26(7):596–615
3. Lauwers B, Klocke F, Klink A, Tekkaya AE, Neugebauer R, Mc Intosh D (2014) Hybrid processes in manufacturing. CIRP Ann Manuf Technol 63(2):561–583

4. Chu W-S, Kim C-S, Lee H-T, Choi J-O, Park J-I, Song J-H, Jang K-H, Ahn S-H (2014) Hybrid manufacturing in micro/nano scale: a review. Int J Prec Eng Manuf-Green Technol 1(1):75–92
5. El-Hofy H (2005) Advanced manufacturing processes. Nontraditional and hybrid machining processes. McGraw Hill New York
6. Shrivastava PK, Dubey AK (2013) Electrical discharge machining–based hybrid machining processes: a review. Proc IMechE Part B: J Eng Manuf 228(6): https://doi.org/10.1177/0954405413508939
7. Ruszaj A, Skoczypiec S, Wyszyński D (2017) Recent developments in abrasive hybrid manufacturing processes. Manag Prod Eng Rev 8(2):81–90
8. Klocke F, Soo SL, Karpuschewski B, Webster JA, Novocic D, Elfizy A, Axinte DA, Toenissen S (2015) Abrasive machining of advanced aerospace alloys and composites. CIRP Ann Manuf Technol 64(2):581–604
9. Jawahir IS, Attia H, Biermann D, Duflou J, Klocke F, Meyer D, Newman ST, Pusavec F, Putz M, Rech J, Schulze V (2016) Cryogenic manufacturing processes. CIRP Ann Manuf Technol 65(2):713–736

Chapter 3
Application of Hybrid Machining Processes in Industry

3.1 Role of Hybrid Machining Processes in Sustainable Manufacturing and Production 4.0 Strategy

From the previous considerations it follows that hybrid processes are an important element in modern manufacturing systems. By using a hybrid process, it is possible to significantly increase the capacity and efficiency of manufacturing processes in relation to the capacity of the componential processes. A very good example is the use of a hybrid process combining electrochemical and electrodischarge phenomena (ECDM) [1]. When such a process is used to shape metallic or composite materials on the metal matrix, the workpiece material is removed by dissolution (transition of metal ions of the workpiece material to the solution). This process is assisted by electrical discharges, as a result of which the material is removed from the workpiece; and in many cases, electrical discharges support electrochemical reactions by depassivation of the workpiece surface. It turns out that this process can be applied to the machining of non-conductive ceramics, for example based on Al_2O_3 matrix. In this case, electrochemical processes do not take part in the material removal but create favorable conditions for the occurrence of electrical discharges in areas filled with gases (e.g. hydrogen emitted on the cathode), as a result of which the material is removed from the ceramic workpiece. This process will be described in detail in Chap. 10. According to the authors, it is a "flagship" example that the proper use (synergy) of physico-chemical phenomena creates new opportunities in the area of manufacturing according to the concept and philosophy of sustainable manufacturing and the fourth industrial revolution called—Industry 4.0 (*Industry 4.0*). The strategy of Industry 4.0 was created by combining information technology with computer-controlled machines (CNC) in an on-line network. Details concerning changes in the approach to the development strategy of industrial manufacturing systems and chronology of subsequent stages can be found in numerous thematic studies, e.g. [1, 2].

In order to determine whether an industrial plant applies or implements the INDUSTRY 4.0 strategy (also called intelligent manufacturing), it is necessary to

W. Grzesik and A. Ruszaj, *Hybrid Manufacturing Processes*, Springer Series in Advanced Manufacturing, https://doi.org/10.1007/978-3-030-77107-2_3

check whether machines, technological (production) processes and employees are connected to an intelligent communication network. It is also important to know the extent to which the machinery can be adapted to the needs of the customer. For this purpose, while observing the production process, it is necessary to define how the machine communicates with the workpiece, i.e. whether the machine receives feedback on the next operation from the "workpiece" [3]. The concept of Industry 4.0 is also manifested in such features as: cooperation, virtualization, decentralization, evaluation of possibilities, service orientation, modularity [3].

Initially, a basic feature of sustainable production can be defined as "the recovery and reworking or regeneration of worn out products". It must be ensured that all products are manufactured in an environmentally justified manner—e.g. the material from which the product is made should not only meet operating requirements but should also be "environmentally friendly". In other words, sustainable manufacturing should result from ecological management, it should be connected with ecological design and development of the product and recycling its production elements. That is to say, the product should return to the manufacturer after the end of its lifetime. For example, in the USA the industry of "*second hand cars*" production from parts obtained in car scrapyards emerged. It can be expected that in the future the "product" will end its life at the manufacturer's—i.e. production and operation and disposal will be a closed cycle. In other words, "care for the environment should be part of the company's production culture"[2].

It is widely accepted that 9 pillars of the Industrial Revolution 4.0 can be distinguished [3, 4]. These are:

- autonomous robots integrated into the production process,
- industrial process simulations,
- vertical and horizontal software integration,
- the industrial *Internet of things-IoT*,
- cyber security,
- cloud computing,
- *additive manufacturing* (3D printers)—it is worth noting that in nature, all living organisms are produced additively,
- augmented reality—software and hardware,
- large data sets and their analysis—creating a competitive edge as a result.

The authors of this study believe that the 10th and extremely important pillar of the strategy—Industry 4.0 is BIONICS, which deals with the study of living organisms (plants and animals) from the point of view of the possibility of using the solutions created by the evolutionary nature in technical solutions. Bionics has already found its important place in technical sciences, e.g. as a *bionization of manufacturing*, and an increasing number of scientists and engineers are convinced of the beneficial impact of this field of science on modern civilization. Bionics gives new opportunities to engineers in solving technical problems and reminds that *homo sapiens* is part of the natural environment and only inside it he can rationally develop and use the solutions developed by Nature [1, 5–8].

Currently, some scientists define Bionics "as the application (copying) of biological functions, structures and mechanisms to machine design". Sometimes more general nomenclature like *Biomimetic design* and *Biologically inspired design* is used, which also includes "Bionics" defined above. Bionics can be a very important pillar of the Industry 4.0 concept, if only because all living organisms are built up additively in nature [4, 9].

An important issue is how hybrid manufacturing methods can be integrated into Industry 4.0 taking into account such features as: interoperability, virtualization, decentralization, assessment of capabilities (e.g. manufacturing), service orientation, modularity [3, 4]. It can be assumed that in the future hybrid machine tools (flexible and adaptable) will enable the implementation of many different hybrid processes which are a combination of electrochemical or electro-discharge processing in different kinematic varieties (drilling, cutting, milling) and basic methods of additive manufacturing. Such an extensive hybrid manufacturing platform will enable an increase in the scope of modelling of manufacturing processes and practical implementation of technological and production services—thus fully integrating into the strategy of Industry 4.0.

The use of hybrid manufacturing processes is inextricably linked to the concept of sustainable manufacturing, and especially to the improvement of the efficiency and environmental performance of manufacturing processes, including the subtractive machining of construction materials [1, 2]. In the first case, the use of hybrid machining allows to increase the efficiency of the process and reduce its costs, mainly by shortening the machining time. In the second case, the harmful impact of the process on the environment is reduced, e.g. by using ecological gaseous media instead of traditional machining fluids, and the energy balance (energy consumption of the process) is improved. With regard to methods using additive shaping, the aspect of sustainability can be seen in high material/raw material efficiency (resulting in smaller supplies), high process reliability, high adaptability of manufacturing chains and longer product life. These issues will be the subject of detailed consideration and analysis in Chaps. 7 and 12 respectively.

3.2 Application Areas of Hybrid Machining Processes in Various Industry Sectors

The current global expectations from manufacturing processes come down to increasing flexibility and efficiency/productivity and, on the other hand, to maintaining high quality [1]. In the case of objects with a complex shape, the manufacturing cycle may include several stages implemented on different manufacturing machines (Fig. 3.1). In such cases, the subsequent handling and positioning of parts is usually inefficient due to the loss of time and increased risk of machining errors, which usually results in higher costs of maintaining the required quality (or obtaining lower quality in the worst-case scenario). The problem of larger space needed to

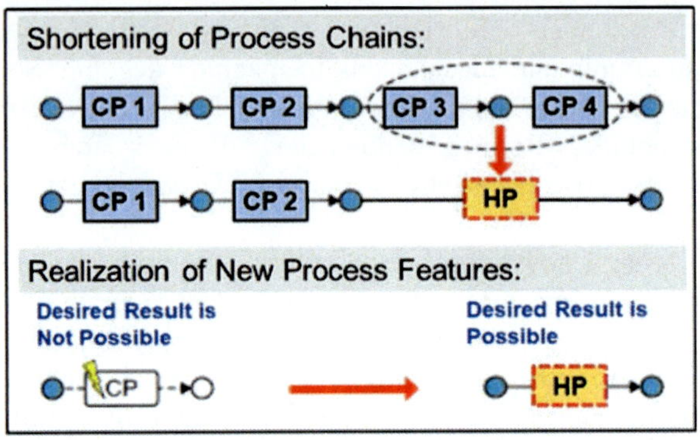

Fig. 3.1 Scheme of process chain shortening and functionality improving resulting from the effect of process hybridization [11]. Symbols: CP-conventional production process, HP-hybrid production process. (www.triz-journal.com)

accommodate many pieces of manufacturing equipment cannot be neglected either. For these reasons, for several decades, design and technological tasks have been taken to integrate/combine multiple processes in a single manufacturing machine, called *a hybrid manufacturing platform* [10]. An example of the first solutions in this area are the CNC *multi-tasking machine tools* for integrated (complete) machining, which have resulted in a significant reduction of costs and machining time. The next steps in the development of integration of manufacturing processes were machines equipped with laser to support thermally conventional machining. The hybrid manufacturing process concept, which is currently being developed on a large scale, is the integration of multiaxial subtractive and additive machining (AM) in one machine [10]. Here, it is possible to use AM machining not only to shape additional workpiece elements, but also to repair expensive, damaged/worn parts, e.g. turbine blades with cavitation pits or surface cracks.

It is important to note that the performance of conventional process can be significantly improved by introducing additional external energy sources. Another possibility is an *effect-cause* analysis for all components of the technological chain leading to an improvement of the whole process, either by integration of single/unit processes or by the cumulative effect of several processes in one hybrid process [11]. These problems are discussed in detail in Chap. 12.

The experience so far shows that hybrid machining processes can be effectively used to fabricate various types of machine parts and tooling, electronic devices and miniature parts made of high-tech materials that require high accuracy and surface quality, high productivity and advanced automation. It should be emphasized that modern construction materials, such as composites reinforced with ceramic and carbon fibers, super alloys and ceramic materials, provide very good thermal, physical, chemical and mechanical properties (including high strength, high *mass-to-volume ratio* and increased resistance to corrosion and wear). Usually these properties do not correlate with good machinability and, in many cases, the machining of these materials using conventional methods is difficult and uneconomical.

Today's industry is producing more and more components with complex 3D configurations, narrow dimensional tolerances and different functional limitations during the design phase, which creates a requirement to search for new machining methods and ways to improve its cost-effectiveness, accuracy and often environmental performance. The basic challenge facing hybrid machining is, as already indicated, a synergistic combination of componential processes to overcome individual limitations and achieve an efficient subtractive process (Fig. 3.1).

The main problems that are solved by hybrid machining include [1, 12, 13]:

- rapid development and growing application in key and strategic industries of difficult to machine materials (DTM), such as automotive industry, aerospace industry, tooling industry, mechatronics industry, nuclear industry, biomedical industry, electronic industry, etc.
- manufacturing miniature structures such as MEMS and NEMS.
- precision machining of parts with complex shape and/or challenging size (so-called artefacts),
- micro and nano machining.
- machining in inaccessible places,
- different requirements in the machining of holes, i.e. shaping of non-circular holes, deep holes, micro-holes, large number of holes located in close proximity (e.g. in tube sheets of micro-heat exchangers),
- machining of structures with low rigidity.

Conventional (mechanical) machining methods face significant limitations due to the strength/rigidity of tools and machine elements, and also due to the chip removal mechanism itself, which can lead to a large plastic deformation and excessive mechanical load on the elements of the technological machining system (MFWT—machine tool-fixture-workpiece-tool). In particular, they are not suitable for machining hard and brittle materials. Unconventional methods, on the other hand, have numerous limitations resulting from specific physical phenomena associated with material removal. For example, laser machining (LBM) has limitations concerning the dimensional criterion of the part, electro-discharge machining (EDM) and electrochemical machining (ECM) are limited to electrically conductive materials, precision WJM machining requires a sufficiently powerful pump to produce the water jet. Therefore, the hybridization of subtractive machining processes using different interactions of mechanical, chemical and thermal influences is a very complex scientific and practical problem. It applies not only to the process itself, but also to the machine tool (thermal stability, accuracy of the feed drives, accuracy of control systems, etc.) and process monitoring. New challenges also concern numerical simulation and optimization of hybrid processes of different structure

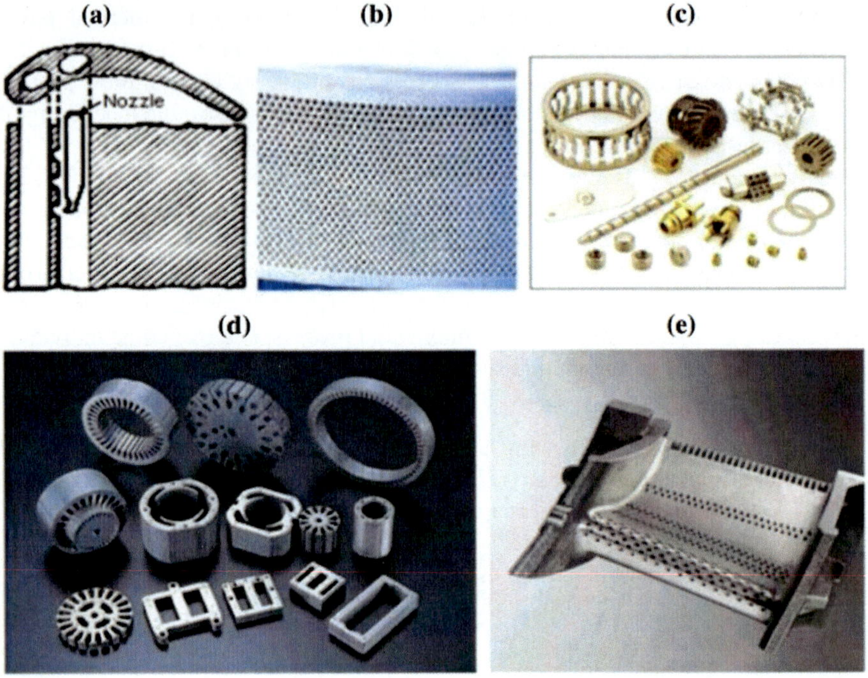

Fig. 3.2 Examples of unique applications of hybrid machining processes: **a** drilling of inaccessible holes (e.g. perpendicular to the bulkhead wall); **b** drilling of a large number of holes on the surface; **c** nano-, micro- and meso-scale machining; **d** machining of parts with complex structural fragments (artifacts); **e** deep-hole drilling on curvilinear surfaces, e.g. cooling holes (channels) in turbine blades [12]. [ww.home.iitk.ac.in]

and scale. Examples of unique applications of hybrid machining to shape complex geometric structures are presented in Fig. 3.2.

Great industrial opportunities have already been achieved through research on hybridization of additive (AM) and subtractive machining. The first stage of such integration, dating back to the late 1980s and early 1990s, consisted of installing a laser source to support conventional machining. The current stage of hybridization of production/machining processes involves carrying out in one machine the additive and final/finishing machining (subtractive) with automatic dimensional control [1, 10]. Additive machining is usually referred to in the literature as "3D printing".

As shown in Fig. 3.3, hybrid machining in this variant is a compromise between the high complexity of the product design characteristic for additive shaping and much higher productivity achieved in CNC machining. It should be noted that the productivity of the AM process is an order of magnitude lower than that of CNC machining and users always have the dilemma of choosing either higher productivity (greater thickness of the material layer applied) or better quality (lower thickness of the material layer). As a result, the hybrid manufacturing process is cost-effective

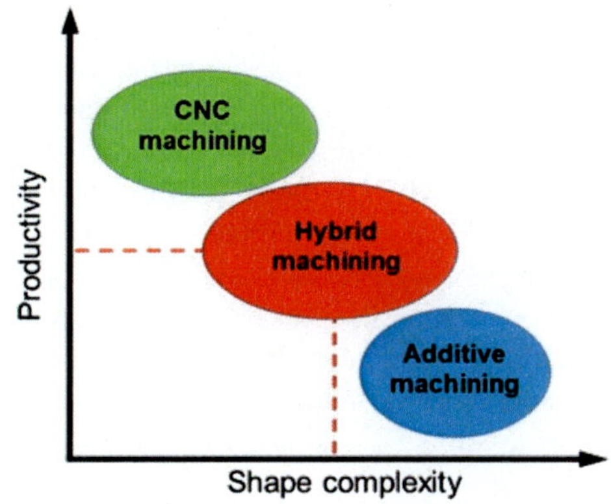

Fig. 3.3 Properties of componential processes and their integration in hybrid machining process of (AM+CNC) type [14]

for smaller and mid-size batches because there is the possibility of the independent monitoring of productivity and surface quality [10].

The range of possible solutions is very wide, among others, it is possible to shape additional elements of the object by 3D printing and powder surfacing (LMD—*laser metal deposition*), cladding elements from materials other than the base material, but also repairing expensive or damaged/used elements. Figure 3.4 shows the elimination of the support structure in the classic 2.5 D-type additive process, in which subsequent layers are deposited in the direction of the axis (+Z). Then the CNC machine tool and the layer deposition device installed on it have additional degrees of freedom and can be controlled in 5 axes. These issues will be described in more detail in Chap. 12.

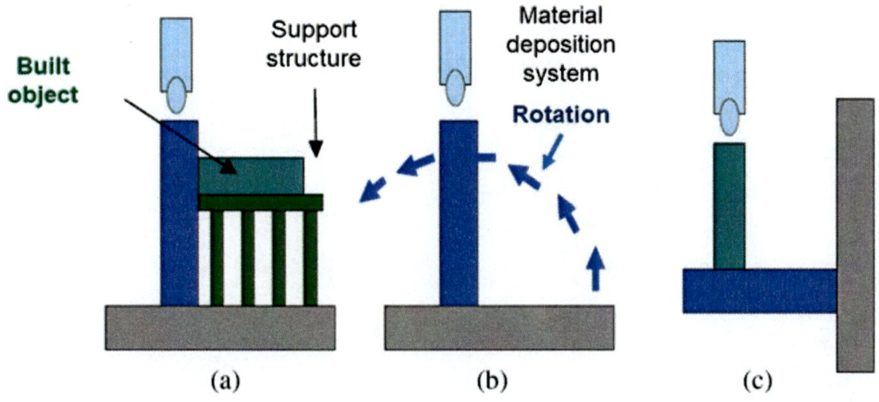

Fig. 3.4 Principle of a manufacturing platform in the AM+CNC hybrid process: **a** additive process using supporting structure; **b** the use of multi-axis control—rotation after column deposition; **c** process continuation in another direction after table rotation [14]

Initiators and users of hybrid and additive manufacturing technologies are industries that use advanced technologies on a large scale, such as the automotive, aerospace and consumer goods industries. Further applications are increasingly emerging in the medical, dental, consumer electronics and tooling industries (mold and die manufacturing). Advanced technology for the additive manufacturing of metallic parts is used to produce a small number of industrial end-use parts, and an increasing number of original equipment manufacturers (OEMs) are using it as a complementary technology and an integral part of their production processes. It is predicted that hybrid/additive manufacturing is a process that will gradually be introduced into many manufacturing industries. It will also apply to micro- and nano-machining processes (Sect. 3.3).

3.3 Applications of Hybrid Micro and Nano-Machining Processes

Micro-manufacturing/machining has now reached a high level of interest from industry because micro-components/products such as micro-displays, micro-sensors, micro-batteries, etc. have already found numerous practical applications in such industrial sectors as: automotive, aerospace, photonic (photoelectronic), renewable energy and medical equipment (Fig. 3.5). They are usually made of many materials, including materials with poor machinability, consist of microstructures with complex

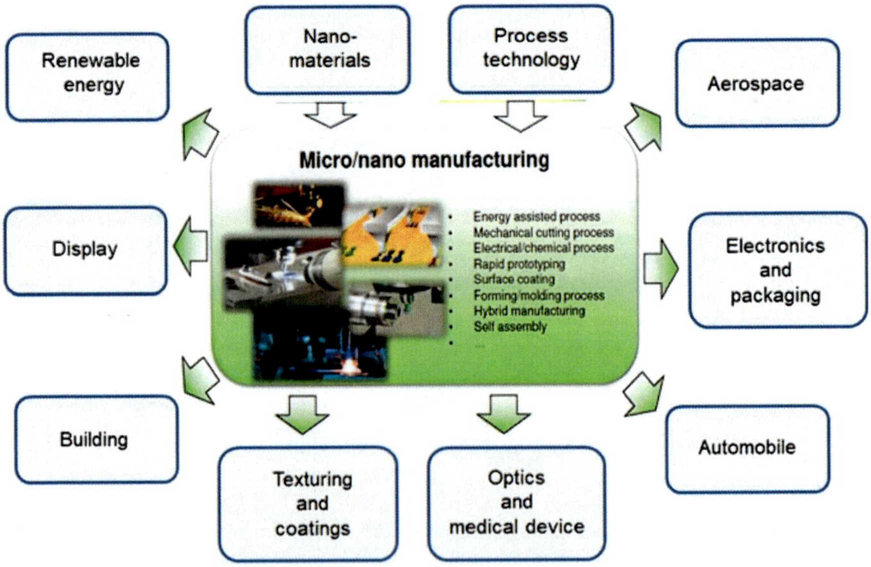

Fig. 3.5 Schematic view of possible applications of micro/nano-scale manufacturing processes in various industrial sectors

shapes and require very high (sub-micron) machining accuracy. For this reason, a large number of hybrid micromachining processes are used in practice, which integrate various conventional and unconventional micromachining processes in order to improve the machinability of the materials used, increase the geometric accuracy of the components, surface quality and performance, and reduce the mechanical load.

In practice, additive and hybrid micro/nano-machining processes are used, as shown in Fig. 3.6. It should be noted that in the case of polymeric materials, the applied subtractive processes are essentially based on the phenomenon of laser ablation. Hybrid processes, which according to CIRP classification (Fig. 2.1) belong to the groups of assisted and combined processes (Fig. 3.7), dominate in the subtractive shaping of metallic materials in microscale.

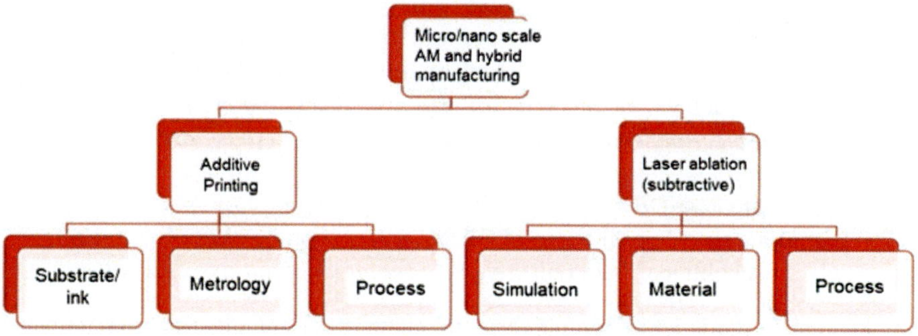

Fig. 3.6 Examples of micro/nano-machining processes applied in industry [15]. (www.imse.iastate.edu)

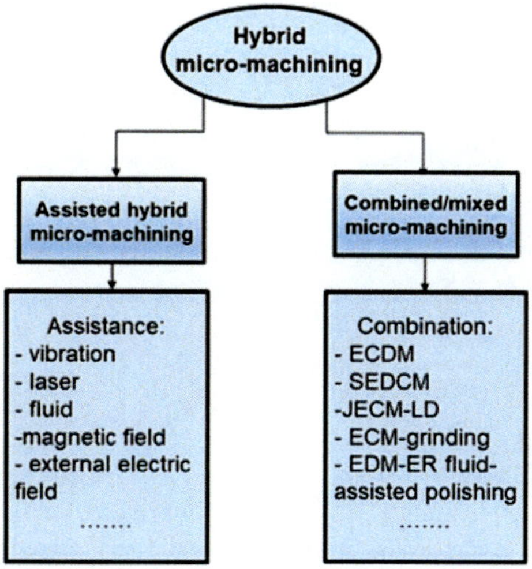

Fig. 3.7 Classification of micro-machining processes [16]

There are two [16] or three [17] micro-machining groups. These are:

1. Assisted processes, which include:
   - processes assisted by the energy of mechanical vibrations superimposed to the tool, workpiece or working fluid.
   - laser-assisted processes, mainly micro-milling and micro-grinding with workpiece heating, electrochemical micro-machining with the location of the cutting zone by laser and water jet with heating and softening of material by laser.
   - processes assisted by liquid, gaseous and chemical media or special fluids that affect the material, such as laser assisted water or methanol jet processing, EDM assisted by magnetic field, external electric field and/or ultrasonic vibration for EDM of carbon fiber reinforced composites.
2. Combined processes, including:
   - micro-ECDM (micro-electrochemical discharge machining) and its modifications such as: ECSM (*electro-chemical discharge (spark) machining*), ECAM (*electrochemical arc machining*) and SACE (*spark-assisted chemical engraving*),
   - simultaneous ED/EC micro-milling and grinding,
   - simultaneous combination of EDM and polishing with the feeding of electrorheological fluid (*ERP-electrorheological fluid-assisted polishing*).
3. Processes carried out with hybrid tools, including multi-function tools or combined tools used in large-scale and mass production.

Figure 3.8 shows the differences in the application structure of hybrid processes in micro- (a) and nano-scale (b).

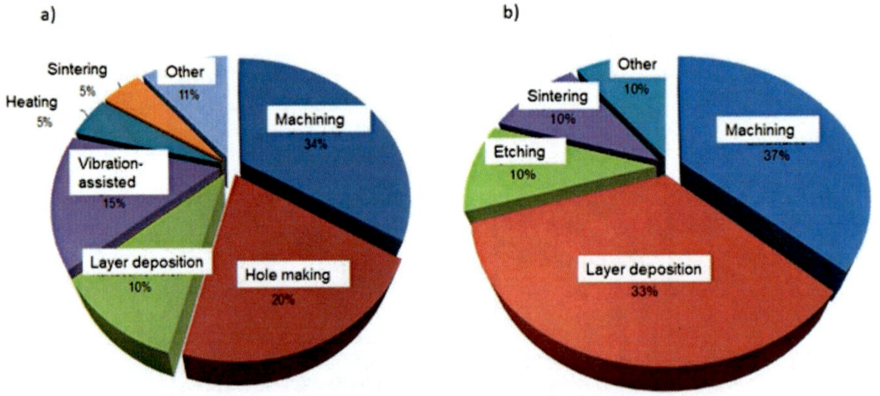

Fig. 3.8 Percentages of componential processes in hybrid manufacturing processes in micro- (**a**) and nano-scale (**b**) [1, 18]

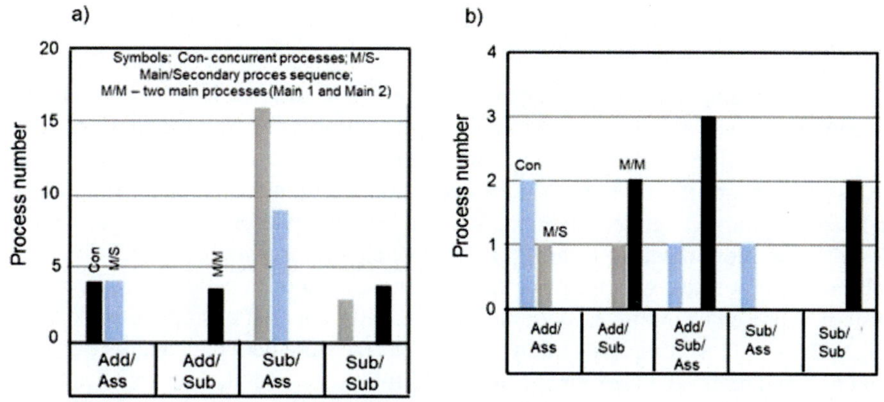

Fig. 3.9 Combination of componential processes in hybrid micro-(**a**) and nano-(**b**) manufacturing [18]. Symbols of process combinations: Add/Ass-additive/assistive, Add/Sub-additive/subtractive, Sub/Ass-subtractive/assistive, Sub/Sub-subtractive/ subtractive, Add/Sub/Ass-additive/subtractive/assistive

Figure 3.8 shows clearly that subtractive and additive processes have the highest percentage in micro-machining, about 64%. On the other hand, in nano-processing, the additive process (layer deposition) is more important (about 33%), despite the still significant use of machining process. Possible combinations of the component processes that make up the final hybrid process in micro- and nano-machining are shown in Fig. 3.9 a and b respectively.

It should be noted that at the current level of development the best results are achieved for a combination of additive and subtractive processes, which can be carried out simultaneously or sequentially (Fig. 3.9). Figure 3.9a shows that S/A processes are most often used in micro-machining—about 56.8%. However, approximately 47.7% of all hybrid micro-processes are carried out as separate processes. In the case of hybrid nano-processes (Fig. 3.9b), almost 30.7% of the total are A/S/S type processes, i.e. with an additional third assistive process. However, approximately 69.2% of these processes are carried out separately.

In principle, there is no limit to the number of componential processes required to produce a 3D component. As mentioned earlier, the problem is to supply the input material and transform it into a final product in one machine tool or workstation. In consequence, the process consists of an initial stage using the AM process and a subsequent stage, which makes optimal use of the capabilities of various assisted processes in order to achieve the highest possible accuracy [19]. Incidentally, all criteria for the implementation of Industry 4.0 are met.

Micro- and nano-scale subtractive machining processes are already the subject of numerous review articles, e.g. [16, 18, 21] and books, e.g. [13, 19, 20].

3.4 Future Vision of Hybrid Manufacturing Processes

The future development of hybrid manufacturing on a macro-/micro- /nano- scale is essentially seen as a process of running the entire process on a single machine tool or workstation so as to obtain the final product regardless of the manufacturing steps necessary to shape it [20]. For this purpose, various computer-controlled componential processes are integrated on a single manufacturing platform (this issue will be described in more detail in Chap. 12). The principle of integration of additive and subtractive processes, which can be realized in an open or in a closed loop control system, is presented in Fig. 3.10. In this case, 12 different relationships can be distinguished, i.e.: (1) next material addition, (2) next measurement data acquisition, (3) next material removal, (4) exchange of the additive process for the subjective process without verification of the result of material growth, (5) exchange of the subtractive process for the additive process without verification of material removal, (6) verification of material growth results, (7) verification of material removal results, (8) additive process with additional reference to the previous measurement and characteristics, (9) subtractive process with additional reference to the previous measurement and characteristics, (10) verification of the process completeness on the basis of the measurement or characteristics of the finished part, (11) non-verified (in an open

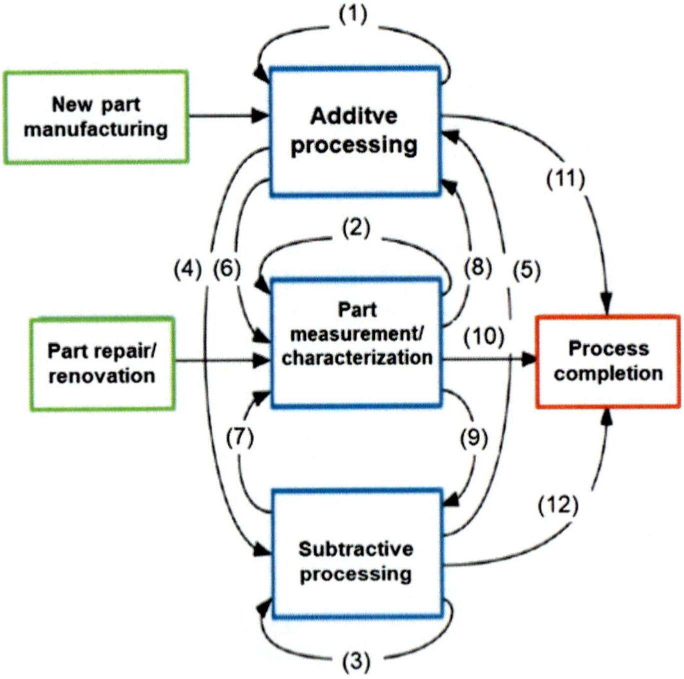

Fig. 3.10 Scheme of interactions in hybrid additive and subtractive processes including operations in open and close-loop systems [19]

loop) completeness of the process after the end of the additive process, (12) non-verified (in an open loop) completeness of the process after the end of the subtractive process. It should be added that the manufacturing of a new part starts with the addition of material in the additive process. On the other hand, the repair or renewal of parts involves the necessary measurement or characterization of defects in order to determine the position or orientation on a 3D printer/machine.

The presented information shows that in the future, the development of structural elements of hybrid manufacturing platforms is expected in the general architecture of the reconfigurable system, control systems, process planning software, metrology and further integration of additive and subtractive processes [19]. The last-mentioned task will be better realized by a holistic consideration of the requirements the two componential processes. An important task is to carry out all the processes during the shaping of the parts in the feedback loop system. In order to prevent mixing and to achieve sustainable recycling/removal of the remaining microparticles from additive machining and cutting chips, the use of management waste material (*swarf management)* should be considered.

The basic micro-machining techniques used effectively for shaping 3D microstructures are shown in Fig. 3.7. The respective percentages of research investigations on hybrid assisted and combined processes are given in Fig. 3.11. The majority of the research concerns assisted processes (about 70%), while the remaining 30% of research concerns combined processes. In the group of assisted processes, processes (about 58%) with additional influence of vibrations dominate, which is related not only to the simple implementation of such a process, but also to a significant improvement of its course. Next, with a 21% share, laser-assisted processes are used, including micro-milling and micro-cutting of brittle and hard materials. Additionally, ultrasonic vibrations, magnetic fields, etc. are used as a secondary source of assistance. In the group of combined processes, the dominant role of the ECDM process with a 67.5% share, which is particularly effective for shaping non-conductive, brittle and hard materials, is noticeable. In turn, the relatively new SEDCM process (15.5%) is considered to be very suitable for shaping 3D structures

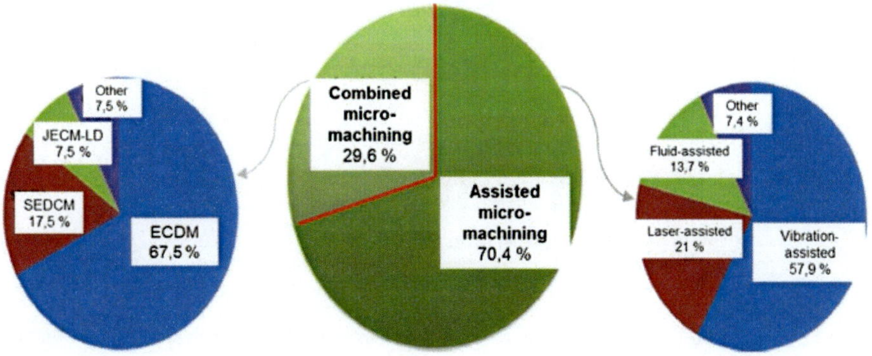

Fig. 3.11 Percentages of research investigations of hybrid micro-machining processes [17]

from conductive materials. The new processes in both groups only account for about 7.5% of the research, which is an obvious challenge for the development of modern micro-machining processes with multiple energy sources [19].

Figure 3.12 shows a promising concept for the hybrid micro-/nano-manufacturing of complex MEMS/NEMS components carried out on a single device or workstation regardless of the number of steps required to complete a 3D element. In this solution, the support structure is first deposited and then machining operations are performed to ensure the required mold shape for polymer injection. In the second workstation, an ideal outline for the placement of the part is made by laser operation and electronic connections are made in the printed circuit between the deposited elements (such as controllers, actuators, communication module, power sources). In the last workstation, final operations such as finishing and removal of the support are performed, after which the product is ready for use.

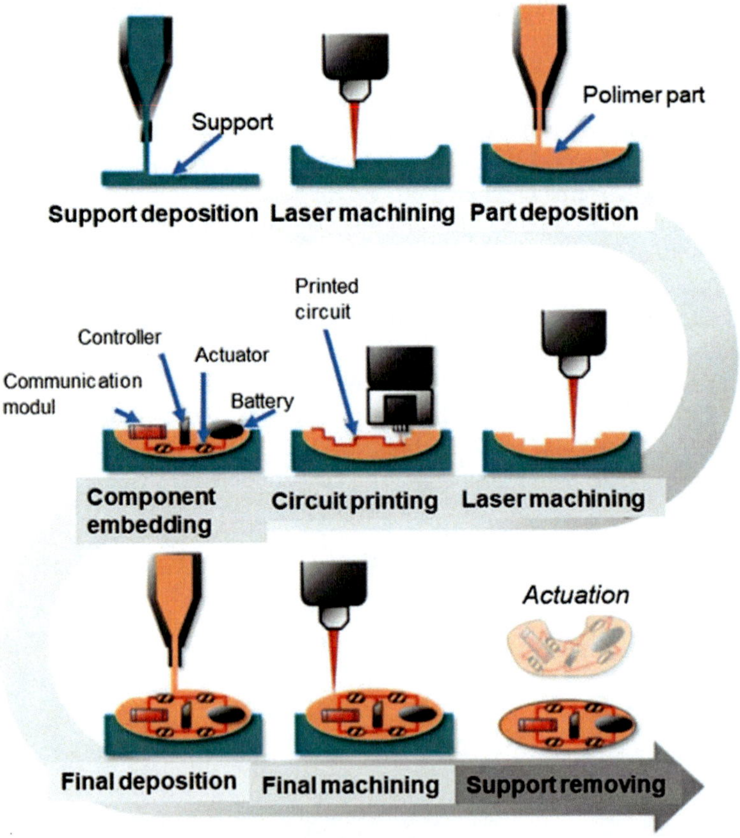

Fig. 3.12 Schematic illustration of integration of future manufacturing processes in micro-/nano-scale [18]

With the development of manufacturing techniques and methods, research into surface quality, dimensional and shape accuracy and performance is essential, in line with the general principle that a hybrid process is by nature more efficient than individual componential processes. Predicted further possibilities include the combination of multiple materials, multifunctionality, matching new special materials to the required properties and achieving ultra-accuracy of produced parts.

References

1. Lauwers B, Klocke F, Klink A, Tekkaya AE, Neugebauer R, Mc Intosh D (2014) Hybrid processes in manufacturing. CIRP Ann Manuf Technol 63(2):561–583
2. Grzesik W, Niesłony P, Kiszka P (2020) Programming of CNC machine tools (in Polish) PWN, Warsaw
3. Digital Transformation Monitor (2017) Germany: Industry 4.0. www.ec.europa.eu/DTM
4. Kozak J (2018) Mathematical modelling of advanced manufacturing processes. Science library of the institute of aviation no. 56, Warsaw
5. Lothar W, Isenmann R, Moehrle MG (2011) Bionic in patents—semantic-based analysis for the exploitation of bionic principles in patents. Proc Eng 9:620–632
6. Luriie-Luke E (2014) Product and technology innovation: what can biomimicry inspire. Biotechnol Adv 32:1494–1505
7. Davim JP(ed) (2012) Machining of complex sculptured surfaces. Springer. London
8. Steinbuch R, Gekeler S (2016) Bionic optimization in structural design. Springer, Berlin
9. Fassi I, Shipley D (ed) (2017) Micro-manufacturing technologies and their applications. A theoretical and practical guide. Springer International Publishing Switzerland
10. Grzesik W (2017) Advanced machining processes of metallic materials. Elsevier, Amsterdam
11. Roderburg A, Klocke F, Zeppenfield Ch (2009) Design methodology for hybrid production processes. TRIZ J. www.triz-journal.com/archives/2009/04/02/
12. Gupta K, Jain NK, Laubscher RF (2016) Hybrid machining processes. Perspective on machining and finishing. Springer, Heidelberg
13. Luo X, Qin Y (2018) Hybrid machining. Theory, methods, and case studies. Elsevier, London
14. Liou F, Slattery K, Kinsella M, Newkirk J, Chou H-N, Landers R (2007) Applications of a hybrid manufacturing process for fabrications and repair of metallic structures. Rapid Prot J 13:236–244
15. The flexible electronics and additive printing laboratory (FEAP). www.imse.iastate.edu/feap
16. Chavoshi SZ, Luo X (2015) Hybrid micro-machining processes: a review. Prec Eng 41:1–23
17. Unune DR, Mali HS (2015) Current status and applications of hybrid micro-machining processes: a review. Proc IMechE Part B: J Eng Manuf 229(10):1681–1693
18. Chu W-S, Kim C-S, Lee H-T, Choi J-O, Park J-I, Song J-H, Jang K-H, Ahn S-H (2014) Hybrid manufacturing in micro/nano scale: a review. Int J Prec Eng Manuf Green Technol 1(1):75–92
19. Flynn JM, Shokrani A, Newman ST, Dokia V (2016) Hybrid additive and subtractive machine tools-research and industrial developments. Int J Mach Tools Manuf 101:79–101
20. Jain VK (2013) Micromanufacturing processes. CRC Press, Boca Raton
21. Zhu Z, Dhokia VG, Nassehi A, Newman ST (2013) A review of hybrid manufacturing processes. Int J Comp Integr Manuf 26(7):596–615

Chapter 4
Physical Fundamentals of Conventional and Unconventional Machining Processes

4.1 Physical Phenomena in the Zone of Machining Processes

In conventional subtractive machining—cutting and abrasive machining processes, the material removal is performed by overcoming the cohesion forces of the material due to various decohesion (material separation) mechanisms, particularly ductile (plastic) fracture, partially ductile fracture and brittle fracture [1]. In conventional and unconventional subtractive machining, the material removal is the result of four different mechanisms including: impact erosion (jet-impact erosion) occurring due to the impact of a jet of particles on the workpiece material, chemical dissolution, electrochemical dissolution and electro-erosion, melting and evaporation, and ion jet bombardment. Matching material removal mechanisms with non-traditional machining processes is given in Table 4.1. In hybrid machining processes, where different primary mechanisms (energy sources) found in conventional and unconventional machining processes are combined, it is important to understand them and combine them effectively according to the principle of synergy.

4.1.1 Cutting Processes

Cutting process is carried out with a wedge-shaped cutting edge that separates a defined layer of material by applying mechanical energy which causes elastic and plastic deformation of the removed material and its transformation into strongly deformed chips. The tool used has a defined number of cutting edges (one or many) and a defined stereometry.

A simplified scheme of the cutting edge impact on the workpiece material, in which three main zones of interaction are distinguished, is shown in Fig. 4.1. These

W. Grzesik and A. Ruszaj, *Hybrid Manufacturing Processes*, Springer Series in Advanced Manufacturing, https://doi.org/10.1007/978-3-030-77107-2_4

Table 4.1 Mechanisms of material removal for different unconventional/nontraditional machining processes (NTM)

Unconventional/nontraditional processes (NTM)	Main mechanisms of material removal
Abrasive Jet Machining (AJM) Water Jet Machining (WJM) Abrasive Water Jet Machining (AWJM) Ultrasonic Machining (USM)	Impact erosion
Chemical Machining (CHM) Photochemical Machining (PCM) Electro-Chemical Machining (ECM) Electro-Chemical Grinding (ECG) Electro-Chemical Deburring (ECD)	Dissolution
Electro Discharge Machining (EDM) Laser Beam Machining (LBM) Electron Beam Machining (EBM) Plasma Arc Machining (PAM)	Melting and evaporation
Ion Beam Machining (IBM)	Ion bombardment

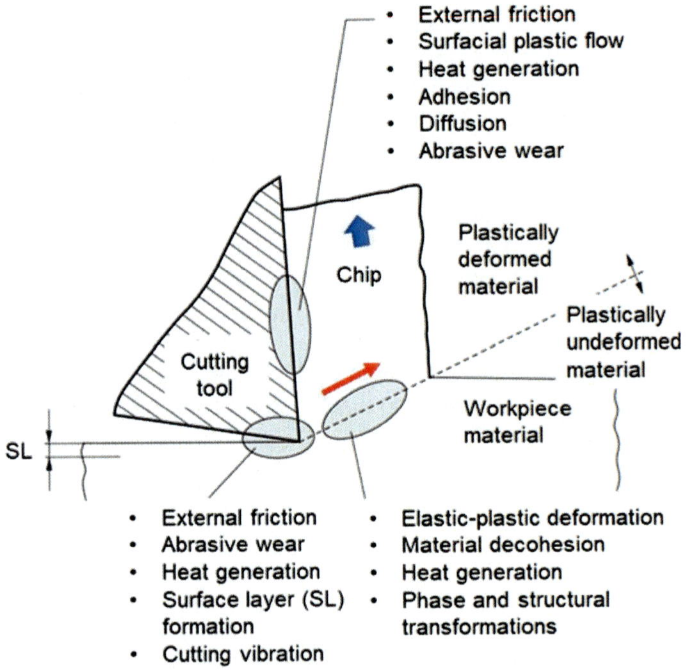

Fig. 4.1 Localization of zones of intensive physical interactions during cutting process [1]

include such physical phenomena as elastic–plastic deformation, material decohesion, sliding friction, heat generation, wear, and surface layer formation. The presented model of separation of deformed layer of material in the form of a chip assumes that the material slides along the *shear plane* (P_{sh}), distinguished by a characteristic angle of inclination (Φ), called *shear angle.* A schematic diagram of the chip formation area with characteristic zones 1–5 is shown in Fig. 4.2. The shape of the cutting edge is characterized by the following geometrical features: the rake angle (γ_o), the clearance angle (α_o), the cutting edge radius (r_β). The ratio of the chip thickness (h_{ch}) to the initial thickness of the cut layer (h) is defined by the chip compression ratio (k_h).

In the case of micro-cutting, the process of chip formation for the three characteristic cases shown in Fig. 4.3, depends on the minimum thickness of the cut layer (h_{min}). This is due to the fact that for a smaller thickness of the cut layer, i.e. when $h < h_{min}$ (Fig. 4.3a) only sliding friction will occur causing elastic deformation of the material. When $h \approx h_{min}$ (Fig. 4.3b) there will be a ploughing of the surface by the cutting edge with a characteristic side flash, which results in the spring back (elastic recovery) of the material subjected to elastic–plastic deformation. If the thickness of the uncut layer increases and $h > h_{min}$, the share of elastic deformation is increasingly smaller and the separation of the material in the form of a continuous chip will be initiated (Fig. 4.3c). The value of h_{min} depends on the ratio of the thickness of the cut layer (feed rate) to the cutting edge radius r_β. A higher value of h_{min} will occur when cutting more ductile materials. Under dry or semi-dry friction conditions, it can be assumed that $h_{min} \approx 0.1 r_\beta$. Therefore, it is important in micro-cutting to make a cutting edge with the smallest possible cutting edge radius, i.e. a very sharp edge, such as in PCD tools [1, 3].

Machining, as a primary process, is subject to various forms of hybridization, i.e. assisted processes (heating of the removed layer by laser and plasma, application of vibrations to the tool edge, liquid and gaseous media assistance, e.g. cryogenic media)

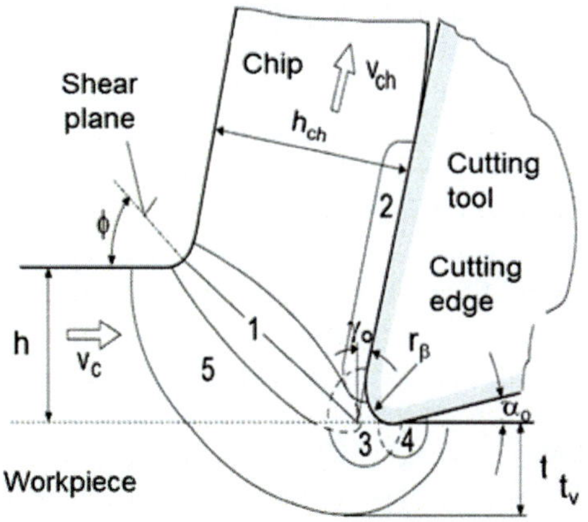

Fig. 4.2 Localization of intensive elastic–plastic deformations in the chip formation zone for orthogonal cutting [1, 2]. Zone symbols: 1—primary deformation zone, 2—secondary deformation zone, 3—zone of penetration of the cutting edge, 4—zone of penetration of the flank face, 5—zone of preliminary deformation at the depth (t_v) beneath the machined surface. *Permission from Springer*

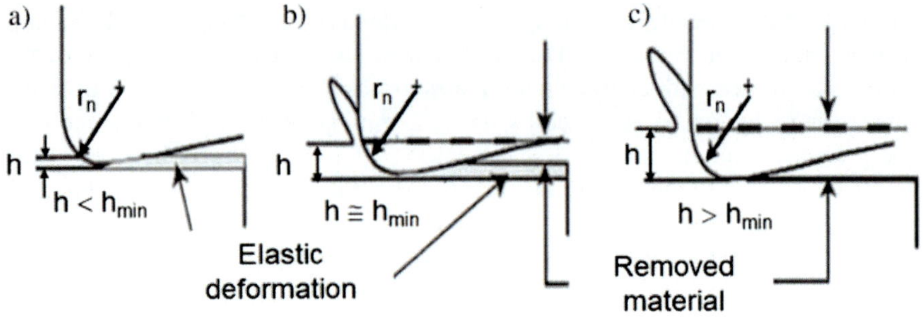

Fig. 4.3 Influence of uncut chip thickness on the chip formation in micro-machining process [1, 3]

and with controlled process mechanism (consisting in simultaneous performing two different cutting operations, e.g. turning and broaching, or milling and turning, or cutting and plastic forming, e.g. turning/milling and burnishing) are used. In consequence, process efficiency and technological quality are improved when controlling the mechanical properties of the machined layer more, e.g., by reduced strain hardening, better chip control, less thermal impact on the tool cutting edge material, reduced diffusive wear of the cutting tool, control of residual stresses in the surface layer, etc. Hybrid machining processes will be discussed in detail in Chaps. 6–11.

4.1.2 Abrasive Processes

The classification of abrasive processes includes processes with bonded abrasive material, i.e. grinding with grinding wheels and abrasive belts, honing and superfinishing, and loose abrasive machining, including such processes as lapping, polishing, ultrasonic machining, abrasive flow machining, massive and vibratory finishing, magnetic-abrasive finishing, aero- and abrasive blasting. The third group, which is already considered hybrid machining, includes electrochemical-abrasive machining [4].

Grinding is a method of abrasive machining, which involves material removal by means of abrasive grains with undefined geometry and not strictly defined their number, which are bonded together in an abrasive tool called a grinding wheel [2]. In grinding, the characteristic of the abrasive tool impact is related to a single abrasive grain of defined stereometry, which represents the active surface of the entire abrasive tool [2, 4]. The actual shape of the cutting edge of an abrasive grain can be determined statistically, but is usually replaced by an approximate spherical, ellipsoidal, or conical shape [4]. As can be seen in Fig. 4.4, a characteristic feature of abrasive grain engagement is the occurrence of a very large negative rake angle, up to (−70°)–(−80°), which does not occur in classical machining [1]. For example, for a grain penetration depth h = 0.25 ρ (where ρ is the cutting-edge radius of a spherical grain), the normal rake angle is equal to $\gamma_n = -48.59^\circ$ [4].

Figures 4.4a and b show the chip formation models for grinding of ductile and brittle materials respectively. For *ductile materials,* the successive zones of contact between the active abrasive grain and the material (I, II and II in Fig. 4.4a) involve the following material deformation and friction mechanisms:

- Zone I includes elastic deformation caused by *sliding* friction of abrasive grain against material,
- Zone II in which plastic deformation of the material is initiated, i.e. a total elastic–plastic deformation occurs. In addition to external sliding friction, internal friction occurs in this area. The predominant deformation mechanism is *grooving/ploughing.*

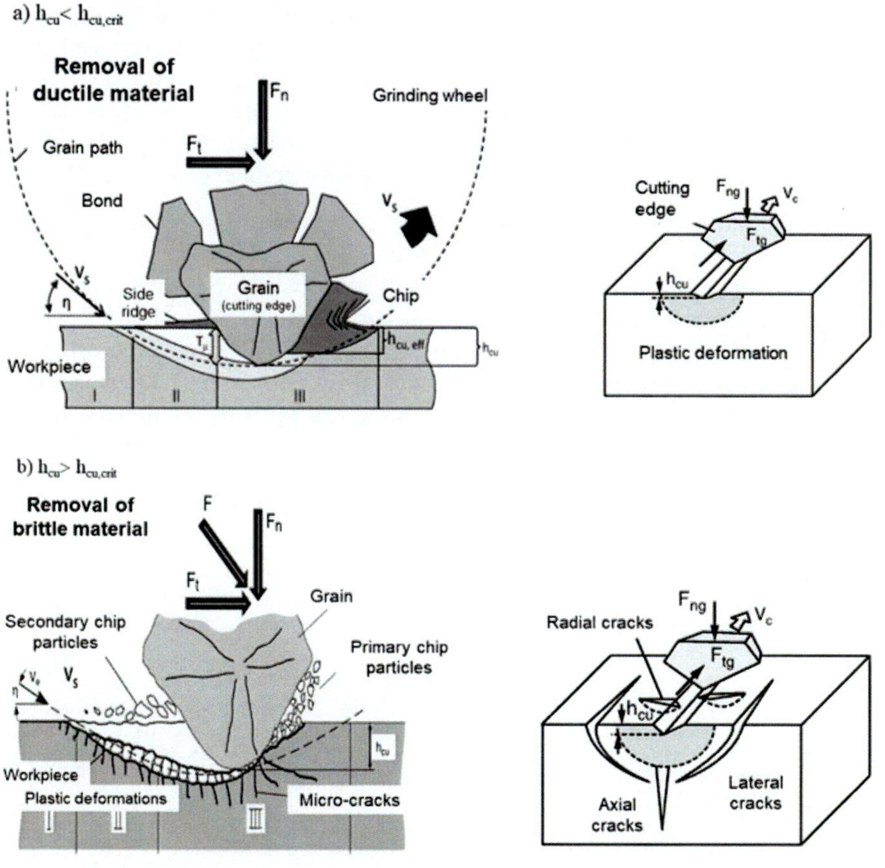

Fig. 4.4 Models of chip formation mechanisms in grinding of ductile **a** and brittle **b** materials [2, 4]. *Permission from Springer*

- Zone III where the micromachining process begins and results in *chip* formation and its separation. This occurs when the penetration of the edge into the material $h > h_{cu,crit}$, and the material begins to flow upwards before the cutting edge. In summary, the material is under, in addition to the previously introduced elastic–plastic deformation and internal friction, an intense abrasive effect of the chip.

For *brittle materials,* the successive contact zones of abrasive grain with the material (I, II and III in Fig. 4.4b) involve the following deformation and friction mechanisms:

- Zone I includes elastic deformations caused by sliding of the abrasive grain against the material,
- Zone II in which residual plastic deformation of the material is initiated, i.e. in total the elastic–plastic deformation occurs. A network of regular micro-cracks is formed in the near-surface layer,
- Zone III in which the micromachining process begins, but as a result of propagation of microcracks in the radial, axial and lateral directions, which finally causes brittle fracture of material particles. That is, chip formation occurs as a result of brittle fracture of such materials, as ceramics or very hard steels and cast irons [1]. An intense abrasive interaction of the chip particle stream in front (*primary chip fractions*) and behind (*secondary chip fractions*) the grain can be observed.

Lapping and abrasive polishing belong to the processes of precision abrasive machining with loose abrasives and differ from grinding with a bonded grain tool (Fig. 4.5a). *Lapping* is a method of machining that consists in removing small particles of workpiece material with the use of loose abrasive powder in which two objects take part (Fig. 4.5b): lapped part and lapping plate, covered with a paste which is a mixture of abrasive powder with fluid (usually an oil-based fluid). Lapping plates with less hardness than the lapped material are used, and as a result the abrasive grains (corundum, carborundum, boron carbide or diamond dust) are partially embedded

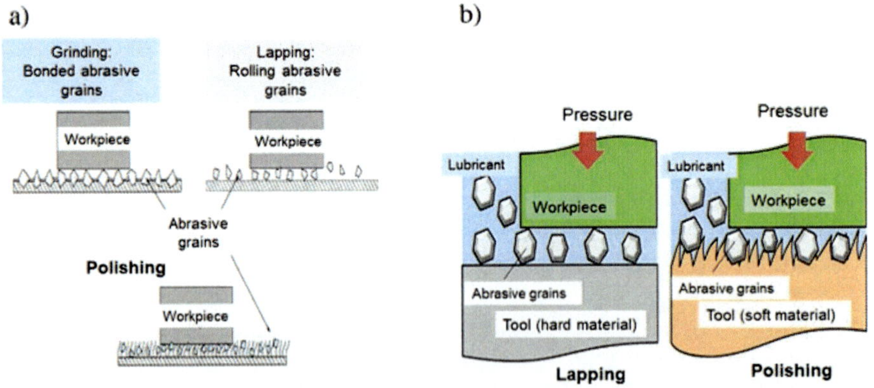

Fig. 4.5 Schematic illustration of material removal mechanisms in grinding, lapping and polishing processes **a** and comparison of abrasive grains' action for lapping and polishing processes **b** [4, 5]

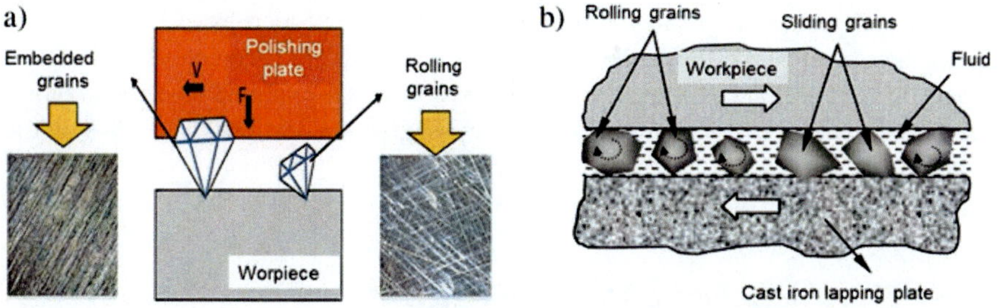

Fig. 4.6 Schematic illustration of material removal mechanisms resulting from abrasion behavior in lapping **a** and polishing **b** [2, 4, 5]

in the lapping surface during the lapping process (Fig. 4.6b) and are used to cut the workpiece with their projecting edges. The lapping plate material, which holds the grains in their temporary positions, plays a role similar to that of a bond in a grinding wheel. In this case, the removal mechanism of material particles is purely mechanical. In the second case (i.e. for chemical–mechanical lapping), which is more commonly used for finish lapping, hard lapping tools and a paste with a soft abrasive, e.g. chromium oxide or iron oxide, are used. The surfactants in the paste produce a thin film of metal oxide on the workpiece surface which can be easily removed with soft grains.

Comparing the work of abrasive micro-grains in conventional lapping and micro-grinding, clear differences can be seen [6]. In lapping, the grain, which is drop-supplied in suspension or in the form of paste, rolls between the lapping tool and the workpiece surface and finds temporary fixation in the lapping tool (Fig. 4.6a). This is the only moment it works in a similar way as fixed grains in the process of micro-grinding with a grinding wheel. The work of abrasive micro-grains also influences the surface topography (Fig. 4.6a) as the unrestrained grains roll around to form grooves (craters) in the form of scratches with material flashes.

Abrasive polishing (more precisely, mechanical-abrasive polishing) uses an abrasive paste with a soft abrasive material applied to the working surface of the tool (Fig. 4.5b). The abrasive tool used is a polishing disc covered or bonded with felt. The aim of classic polishing is to give the workpieces the required surface roughness and gloss. Abrasive polishing does not improve the dimensional accuracy of the workpiece, nor does it eliminate surface shape errors.

Grinding and other abrasive processes, e.g. polishing, as a primary process, are subject to various forms of hybridization, i.e. assisted processes are used (application of vibrations, magnetic and electric fields, support with liquid and gaseous media, e.g. *electro-rheological fluid*, cryogenic media, laser heating, electrochemical dissolution) and with a controlled process mechanism, the so-called grind hardening. In contrast, as an assistive process, grinding is used together with EDM (electrodischarge abrasive, grinding—EDAG) and ECM (electrochemical abrasive grinding—ECAG), while abrasive polishing assists electrochemical processing. Effects increasing process efficiency and technological quality are obtained, such as increased process efficiency, control of mechanical properties of the machined layer. They result in softening hard ceramic material, less thermal and tribological impact on the workpiece material, control of stresses in the surface layer, etc. Hybrid assisted abrasive machining processes will be discussed in detail in Chaps. 6–9, while combined processes are included in Chap. 10.

4.1.3 Electrodischarge Machining (EDM)

In the electrodischarge machining (EDM), the material removal from the workpiece is a result of the phenomena accompanying the electrical discharges occurring in the space between the workpiece and the electrode-tool (Fig. 4.7) [1–4]. This space is filled with a liquid dielectric such as cosmetic kerosene, deionized water, or a gaseous dielectric such as air or gas mixture. The electrode-tool and the workpiece are connected to an electric voltage pulse generator (U = 60 − 120 V,) which initiates electric discharges characterized by a short-term electric current flow ($t \cong 1 - 2000$ μs) of up to 120 A, in a part of the interelectrode gap called the *discharge channel* (Fig. 4.8).

Discharge channels are formed around cathode points where cold electron emission occurs. The electrons are accelerated in an electric field and due to collisions cause local ionization of the dielectric fluid. When the electrons reach the workpiece surface (anode), a discharge channel is fulfilled with a well-conductive plasma. As long as the electric voltage pulse lasting, the electric current flows through this channel and heat is generated causing the temperature of the plasma and electrode surfaces increases to 6000–12,000 K [7, 8]. Therefore, some of the material on the electrode surfaces is vaporized and melted, and the properties of the surface layer are significantly changed in comparison to the raw material. This means that the volume

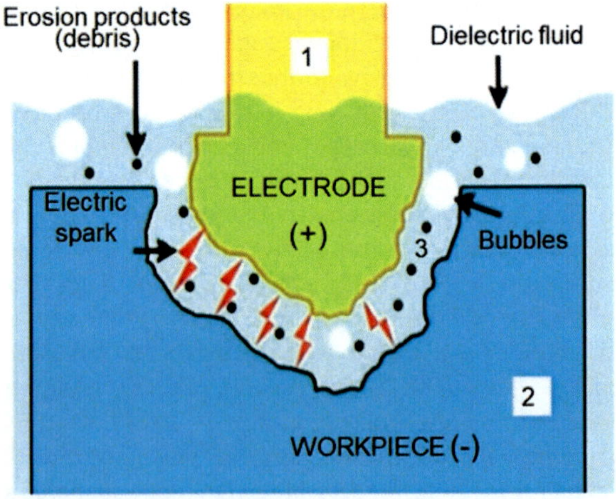

Fig. 4.7 Scheme of EDM sinking: 1—electrode-tool moving in the workpiece direction 2—with the time-dependent velocity, $v_f(t)$, 3—working space with dielectric fluid/gas—cosmetic kerosene, deionized water, air mixture of specially selected gases [7]

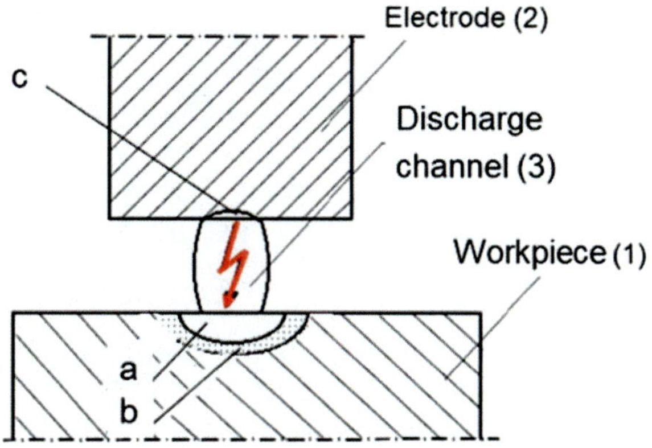

Fig. 4.8 Scheme of the zone of individual discharge: 1—workpiece, 2—electrode-tool, 3—discharge channel which rapidly propagates resulting in removal of melted **a** or evaporated **b** material from the workpiece or electrode-tool **c** [7]

of material removed during a single discharge depends significantly on its melting and evaporation temperature and not on the mechanical properties of the material, such as hardness or strength.

The process parameters are selected so as to minimize the wear of the working electrode for assumed material removal rate. Unfortunately, wear of the working electrode cannot be completely eliminated, which results in a serious technological problem connected with dimensional and shape accuracy [7, 8]. The EDM process described above is used in practice in the kinematic variants of cavities or holes drilling (Figs. 4.7), wire cutting (Fig. 4.9) and milling (Fig. 4.10).

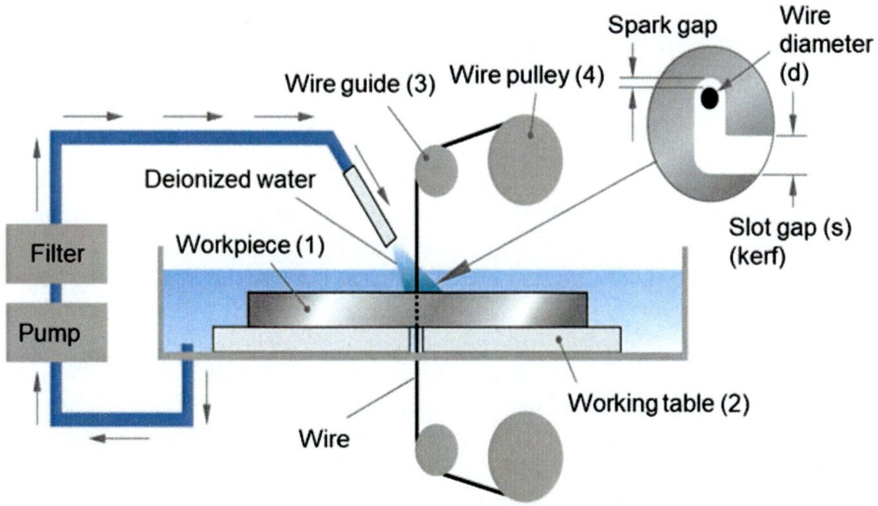

Fig. 4.9 Scheme of wire EDM (WEDM) process; 1—workpiece, 2—working table moving in X and Y directions, 3—electrical connection to the wire, 4—set of rollers for wire rewinding, *s*—thickness of the cutting gap, *d*—wire diameter—typically in the range of 0.1–0.3 mm [7, 8]

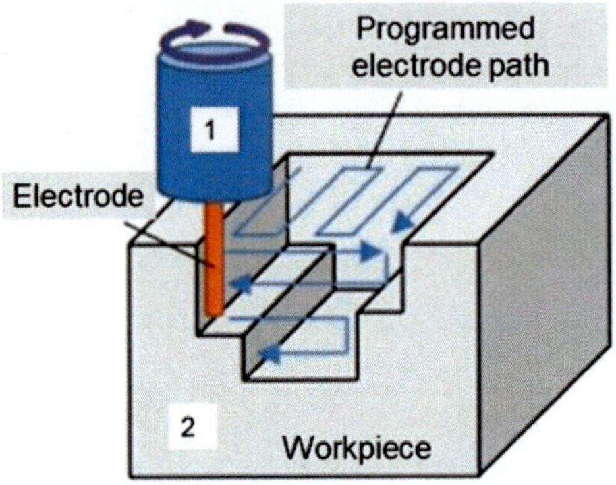

Fig. 4.10 Scheme of EDM milling; 1—rotated universal cylindrical working electrode of typical diameter from about 0.1 mm to several millimetres, 2—workpiece; the shape of machined surface results from the trajectory of working electrode reproduction in machined material [7, 8]

Hybridization of the EDM process, as a main process, may involve the assistance of ultrasonic vibration, laser, magnetic field and special liquid media [7, 9]. As a result, material removal rate increases. Thanks to improved dielectric flow, more efficient removal of material particles from the working gap, increased plasma ionization, better discharge control, less thermal impact on the workpiece surface layer, etc., are obtained. Hybrid EDM processes will be discussed in detail in Chaps. 6, 7, 8 and 9.

4.1.4 Electrochemical Machining (ECM)

Material removal in the ECM process is related to the flow of electric current through the electrolyte flowing through the inter-electrode gap between the workpiece (anode) and the electrode-tool (cathode). The flow of current in the ECM process is associated with the transport of mass and charge by ions across the boundary between workpiece material (anode)—electrolyte. This process follows Arrhenius, Faraday and Ohm's laws (Fig. 4.11) [7, 8]. The flow of electric current through the inter-electrode gap generates large amounts of heat, which increases the electrolyte temperature and can disrupt the processes occurring in the inter-electrode gap. The heat, together with the products of dissolution, must be removed by an intensive flow of electrolyte through the machining area, so that its maximum temperature is about 70–80 °C and the concentration of solution products does not approach a critical value. Under critical conditions, electrical discharge or short-circuit (electrodes mechanical contact) usually occurs, so as local melting or even evaporation of electrode material occurs. As a result, process interruption and electrodes damage occur. To avoid such critical states of the process, the hydrodynamic conditions in the inter-electrode gap should be optimized in the first place. In the case of machining of complex shaped surfaces

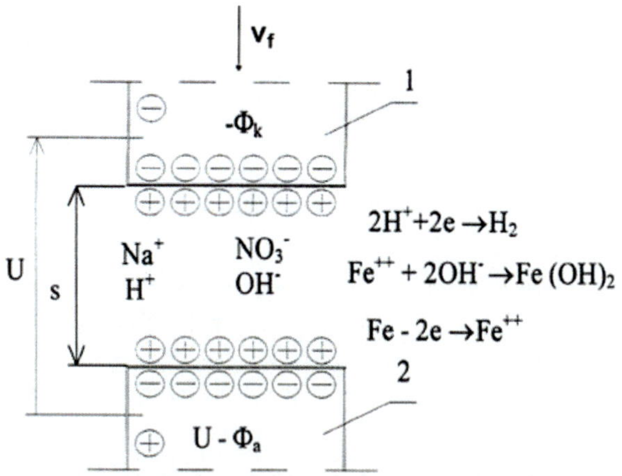

Fig. 4.11 Simplified scheme of inter-electrode zone in ECM process in the case of machining iron element (Fe) using water solution of $NaNO_3$; Φ_a, Φ_k—electrode potentials, basic anodic reaction (material removal): Fe–2e → Fe^{++}, basic cathodic reaction (hydrogen evolution): $2H^+ + 2e \rightarrow H_2$. This means that in cathodic reactions the dimensions of the working electrode do not change during the process and this allows producing a large number of parts, for instance turbine blades of aircraft, using the same working electrode. 1—cathode (electrode-tool), 2—anode [7]

of dies, casting dies, molds or flow engine turbine blades, such a condition may be difficult to reach while maintaining a constant voltage (usually U = 10–30 V). Therefore, the power supply with appropriately selected voltage pulses is often used. In the interval between the voltage pulses, the dissolution process does not take place and the heat and electrode reaction products are removed, so that in the next pulse the dissolution process conditions are optimal [7, 8].

Electrode processes depend mainly on the chemical composition and metallographic structure of the workpiece material, the chemical composition of the electrolyte and the basic process parameters—the interelectrode voltage and the electrode feed rate. Electrode processes are characterized by the so-called *polarization curves* expressing the dependence of the current on the anode potential. The polarization curve of the iron (anode) process is shown in Fig. 4.12 [7]. It allows distinguishing between the active, passive and transpassive states of dissolution. In the active and passive states, the basic reaction is the transfer into solution of iron in the form of divalent ions (Fe^{2+}).

In the passive state, a poorly conductive layer is formed on the anode surface, which disappears at the transition to the transpassive state, in which iron enters the solution in the form of Fe^{3+} ions. Depending on the type of processed material, the basic process parameters, hydrodynamic conditions, and the process state (active, passive, transpassive), phenomena such as etching at grain boundaries, formation of a passive layer, saturation of the surface layer with hydrogen, and cavitation, which usually have a negative influence on the quality of the surface layer, can occur. Information about the distribution of the inter-electrode gap thickness "s" (Fig. 4.11) is

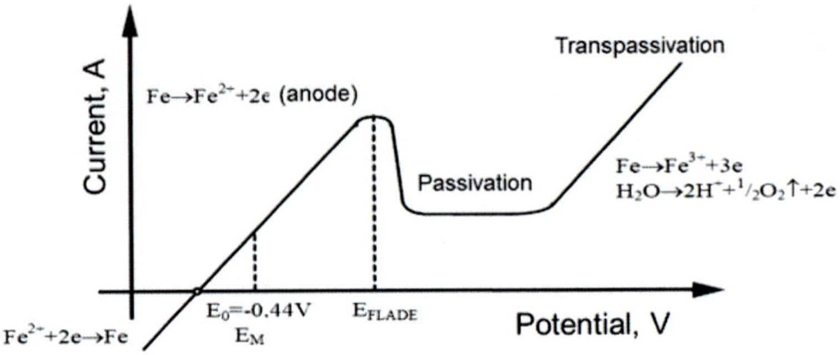

Fig. 4.12 Polarization curve for iron; E_M—metal potential, E_{FLADE}—Flade's potential [7]

necessary to correct the shape of the working electrode. Under typical ECM conditions, the front gap thickness varies from several micrometers to about 0.2 mm and subsequently increases up to 1 mm on wall surfaces parallel to the working electrode feed. The undisputed advantage of ECM is that for optimal and stable process parameters the electrode-tool does not wear. Under typical ECM conditions, the surface roughness varies in the range of $R_a = 0.16\ \mu\text{ m} - 2.5\ \mu\text{m}$ [7, 8].

In industry, ECM is mainly used for manufacturing surfaces with complex shapes with such operations as drilling, sinking and milling. In addition in micro-parts manufacturing operations occasionally wire electrochemical cutting is also applied.

Hybridization of electrochemical machining (ECM) process, as a main process, can refer to the electrodischarge assistance, ultrasonic vibration assistance, laser assistance or magnetic field assistance [7]. As a result, it is possible to improve machining process efficiency, surface layer quality, hydrodynamic conditions of electrolyte flow and removal of machining products from interelectrode space, etc. Hybrid electrochemical machining processes will be discussed consequently in detail in Chaps. 6, 7, 8 and 9.

4.1.5 Water Jet and Abrasive Water Jet Machining Processes

Waterjet machining, especially abrasive water jet machining (AWJM), is widely used for cutting non-metallic materials and hard and difficult-to-machine materials such as titanium alloys, ceramics, metal matrix composites, reinforced concrete, minerals, etc. In WJM machining, the water jet outflows at the velocity of about 900 m/s (about Mach 3), and striking the surface of the workpiece, it acts like a saw producing a narrow groove [10]. In contrast, the AWJM process uses the cumulative impact erosion phenomenon (Table 4.1) caused by the impact of the water jet and abrasive particles at a pressure of 400 -700(800) MPa (4000–7000(8000) bar). It should be noted that the pressure of 380 MPa ensures the continuity of the jet without pressure fluctuations. This is because the mass flow rate (Fig. 4.13b) is the result of

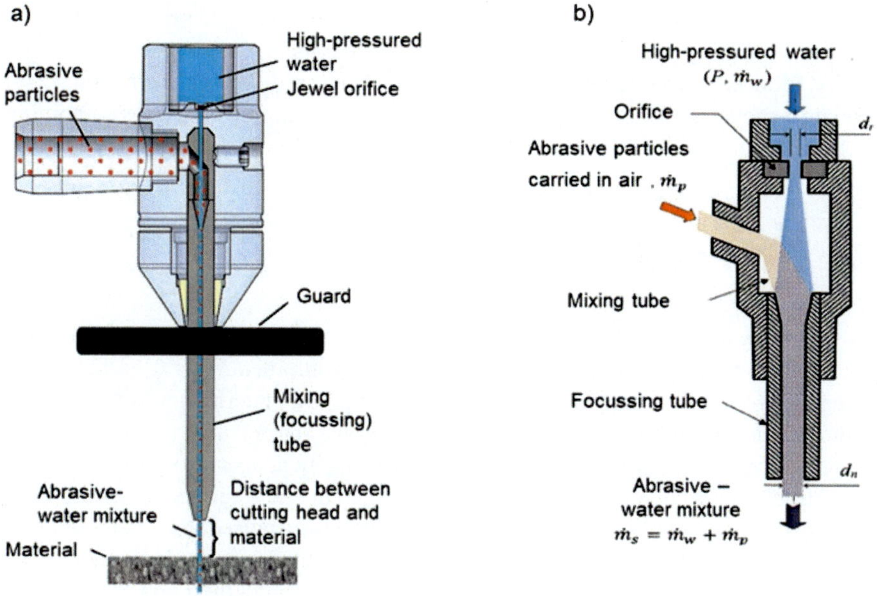

Fig. 4.13 Principle of abrasive water jet machining (AWJM) **a** and definition of jet flow rate **b**

the summation of two components, i.e. water and abrasive particles masses ($\dot{m}_s = \dot{m}_w + \dot{m}_p$). Abrasive waterjet machining is a purely mechanical (cold) process because the temperature rise in the machining zone does not usually exceed only 50–80 °C.

This high pressure is generated by a special intensifier pump (piston) or a crankshaft pump, which is the basic component of the WJM/AWJM installation. Pure water or water-abrasive mixture is delivered through a sapphire or diamond nozzle (Fig. 4.13a). The abrasive particles are delivered from a container and combined with the water jet in a mixing chamber. The main advantage of AWJM is the possibility of precise control of the cutting depth by setting the appropriate distance between the head and the material surface and the direction of jet penetration [7]. Very high cutting efficiency is obtained, for example, cutting a plate of 76 mm (3 inch) thickness made of hard tool steel is performed at about 850 μm/s (2 inch/min). Process parameters that affect productivity and process quality include hydraulic pressure, nozzle-to-surface distance (*stand-off distance*), abrasive flow rate, and abrasive grain material. Important process quality parameters include material removal rate (MRR), surface roughness, width and conicity of the kerf.

The hybridization of waterjet/abrasive waterjet machining (WJM/AWJM) only applies, as a primary and secondary process, to the combination with laser machining [9, 11, 12]. In the first case, in AWJM assisted by the femtosecond or CO_2 laser (*Laser-assisted Waterjet Machining*), the material is heated and softened by the laser and then removed by the waterjet. Melting and evaporation, characteristic of laser machining, do not occur. In the second case (*water-assisted micro-machining*), a water jet surrounds the hole—forming a shield—which is made by a laser beam, and the material is removed by laser ablation. This results in increased process efficiency in micromachining of ceramics, LCD glass and polymers, improved surface layer quality, reduced heat-affected zone (HAZ) and lesser micro-crack formation, more efficient removal of machining products, etc. Hybrid machining processes combining laser machining and WJM/AWJM will be presented in Chap. 9.

4.1.6 Laser Beam Machining Process

Laser beam machining (LBM) uses a laser light beam (Fig. 4.14) as an extremely versatile tool to shape elements of machinery and equipment. It is used for manufacturing by material removal (subtractive shaping), by material addition (additive methods) and for modification of the properties of the surface layer of elements as a result of thermal or thermo-chemical machining in which, apart from the change of microstructure, its chemical composition is changed by melting special materials [7, 10, 13]. The laser light beam emitted by a continuous or pulsed laser is monochromatic with a divergence of the order of $10^{-3} - 10^{-2}$ rad. Thus, the laser light beam can be focused on a surface with a very small diameter and a high power density of $10^{8} - 10^{14}$ W/cm^2 can be achieved. The beam of laser light contacting the workpiece is partially reflected and partially absorbed within a very short time of $10^{-11} - 10^{-10}$ s. The laser beam energy can usually penetrate to a depth of $10^{-6} - 10^{-4}$ mm.

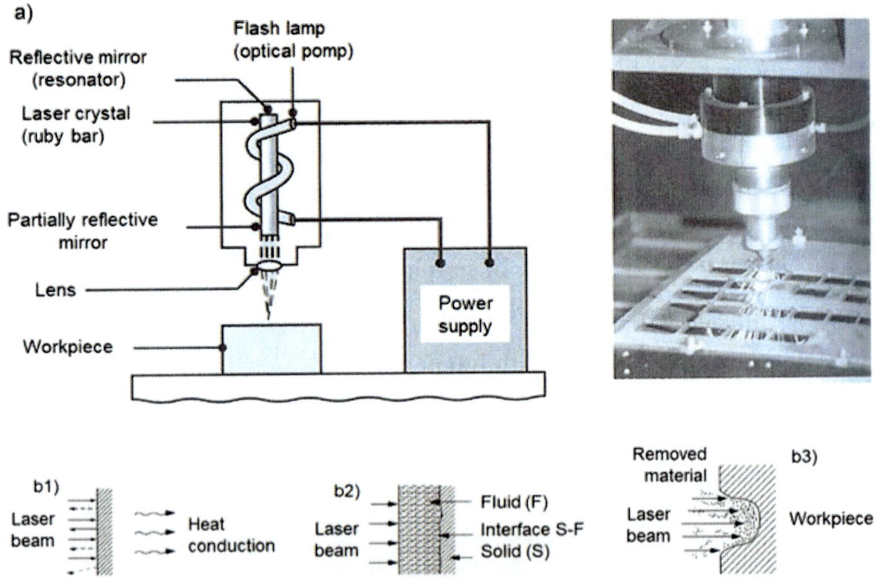

Fig. 4.14 Scheme of laser beam machining **a** and associated mechanisms of material removal (physical phenomena): **b**1 absorption and heating; **b**2 melting; **b**3 evaporation [10, 14]

Temperature stabilization in the penetration volume occurs in the time of $10^{-9} - 10^{-8}$ s. The laser beam energy is converted into thermal energy and causes melting and evaporation of the material. The result of such rapid changes in the material's physical state is the high pressure in the range of 10 - 100 MPa, which causes "explosive" removal of molten and vaporized material. As the material is removed, new layers of material are exposed, which again are penetrated by laser radiation causing their subsequent heating, melting, evaporation and finally removal. It is worth noting that LBM process can be used to shape any conductive and non-conductive materials.

In the case of heat treatment process, the laser light beam heats up the surface of the component, and in the case of thermo-chemical modification, a layer of material, e.g. in the form of powder, is fused into the surface layer of the modified component. Subtractive laser machining can be implemented in cutting operations, deep hole drilling (pits or holes), and in material removal operations equivalent to turning or milling [7, 14]. Many types of lasers are used in laser machining. Relatively most commonly used are CO_2 molecular lasers, solid-state Nd:YAG lasers, and so-called excimer lasers. CO_2 lasers emit radiation at a wavelength of 10.63 μm and can operate continuously or in pulses, and the beam can be focused on a spot area with a radius of 50–500 μm. These lasers are most widely used for laser hollowing, cutting, turning and milling, shaping by heating, assisting the cutting process, and electro-discharge and electrochemical machining.

Solid-state Nd:YAG lasers with a wavelength of emitted radiation of 1.064 μm can also operate continuously or in pulses. The beam emitted by these lasers can be focused on a surface with a radius of 5–50 μm. Due to the much smaller spot diameter, these lasers are used for analogous but more precise operations as with CO_2 lasers, including the manufacture of microelements. Excimer lasers are pulsed gas lasers, in which the active medium is a mixture of noble gas (Ar, Kr, Xe) and halogen gas (F_2 or HCl). These lasers emit radiation with a wavelength in the range of 0.17–0.351 μm. The beam of this radiation can be focused on a surface with a radius less than 1–10 μm. For this reason, they are often used in the fabrication of microelements or microstructures (Fig. 4.15).

Hybridization of the LBM process as a primary process is sometimes accomplished by assisting with magnetic fields or ultrasonic vibrations. On the other hand, laser is widely used to support typical machining processes like turning and milling (LAM- *Laser-assisted machining*), grinding (LAG- *Laser-assisted grinding*), electrochemical machining *(LAJECM- Laser-assisted jet electrochemical machining),* electro-discharge machining (LAEDM- *Laser-assisted electrodischarge machining*) or WJM machining mentioned in Sect. 4.1.5 [7, 10, 14, 15]. The hybridization of the LBM process enables improving the efficiency of the primary process, e.g. by material softening or intensification of material removal in EDM and ECM machining, and thus increasing the efficiency and accuracy of machining. The assistance of LBM for other machining processes will be discussed in Chap. 9.

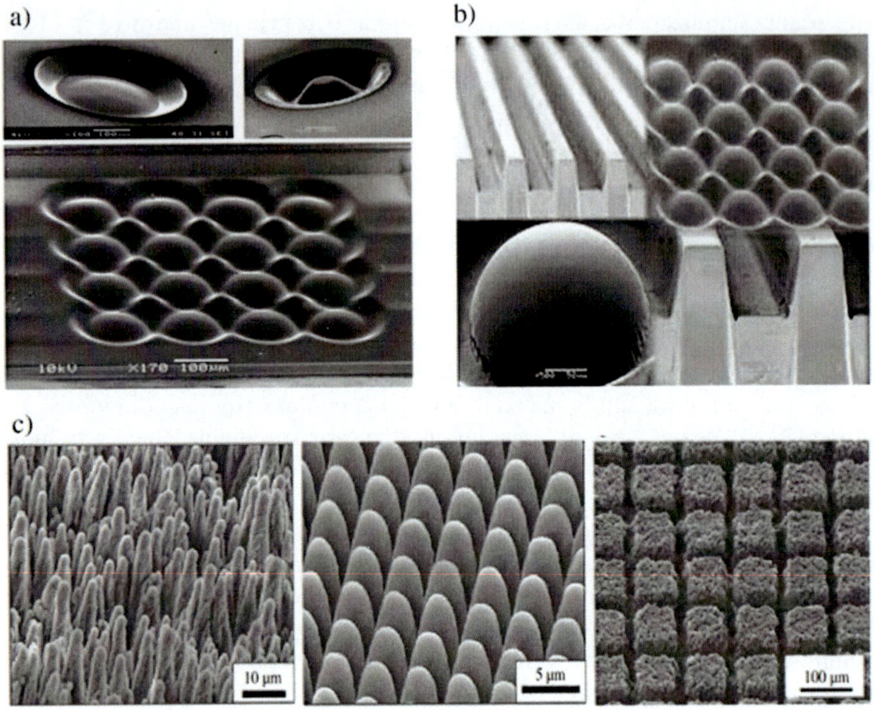

Fig. 4.15 Example of microstructures made by laser: **a** micro-lenses, **b** microelements applied in photonics, **c** microelectrodes [13]

4.1.7 Ion Beam and Electron Beam Machining Processes

Electron beam machining (EBM) is the removal material process performed by melting and evaporation resulting from local heating by an electron beam. The process (Fig. 4.16) is carried out in a vacuum chamber under a pressure of 10^{-4} torr. The electron source is a tungsten cathode heated to a temperature of about 2500–3000 °C [10, 14]. The emitted electrons are accelerated at high voltage—even up to 150 kV. It is estimated that the velocity of electrons reaches about 100–200 km/s. After acceleration, the electron beam is focused by an "electronic or magnetic lens" onto the surface of the workpiece, e.g. with a spot diameter of 0.25 mm. Upon contact with the workpiece surface, the kinetic energy of the electrons is converted into heat which causes the material layer to heat up above the evaporation temperature (the power density is as high as about 1.55 MW/mm^2). In order to prevent heating of the entire workpiece, but only an area of the thin machining zone, a pulsed electron beam with a frequency of about 100 Hz is used, which penetrates to a certain depth into the workpiece material and causes a rapid increase in temperature and an associated rapid increase in the volume of the heated material, followed by its "explosive" removal. The electrons then heat up the deeper layers and also cause their intensive

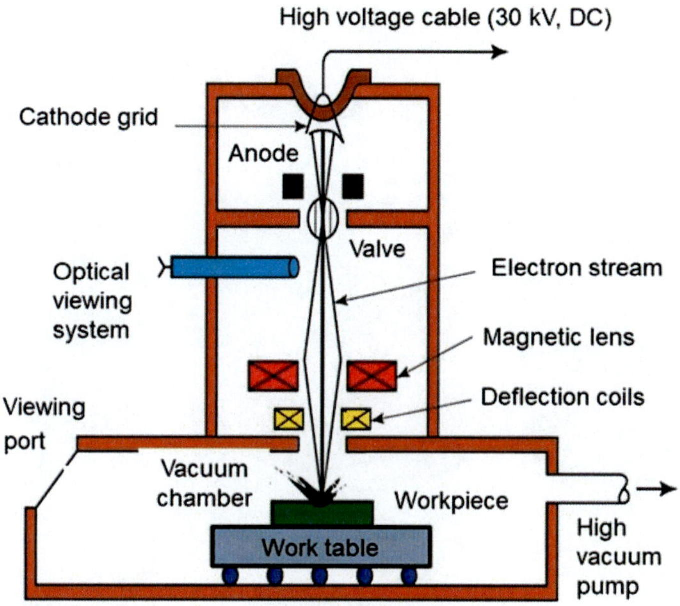

Fig. 4.16 Scheme of an electron gun for electron beam machining (EBM) [15]

removal. In this way, the material around the electron beam is removed at a rate of up to 10 mm^3/min.

Common applications of EBM machining, shown in Fig. 4.17, include hollowing and hole drilling, perforation of thin sheets (up to 5000 holes/s the possibility extended to 600 holes on an 1 mm^2 area.) The time to make one hole is about 10 μs [7, 14]. Further applications include making shaped slots or indentations and many operations in the production of electronic devices, microelements or microstructures of aerospace components [10, 16, 17].

Electron beam melting (EBM) is an emerging modern additive manufacturing process used to produce 3D parts from materials that are difficult to shape conventionally, such as titanium alloys (γ-titanium aluminides γ-TiAl). The main limitation of the application of this process is relatively high roughness of the machined surface, about Ra = 30 μm. This fact forces the use of an additional finishing operation, such as milling, which improves surface roughness to about Ra = 0.1 μm [17].

Ion beam machining (IBM) is performed in a vacuum chamber, and the ion beam is directed to the workpiece—anode by an appropriately selected voltage (Fig. 4.18). The mechanism of material removal is different than in EBM or LBM machining and consists in the removal of atoms by ions bombarding the surface of the anode [14]. Therefore, the process is called ion etching and can be used in milling or ion beam polishing operations. In the device shown in Fig. 4.18a, the ion beam is accelerated by an appropriately selected voltage towards the workpiece, and a lens system focuses it on a selected part of the workpiece surface. The kinematic system

Fig. 4.17 A variety of parts made partly using electron beam machining process [16]

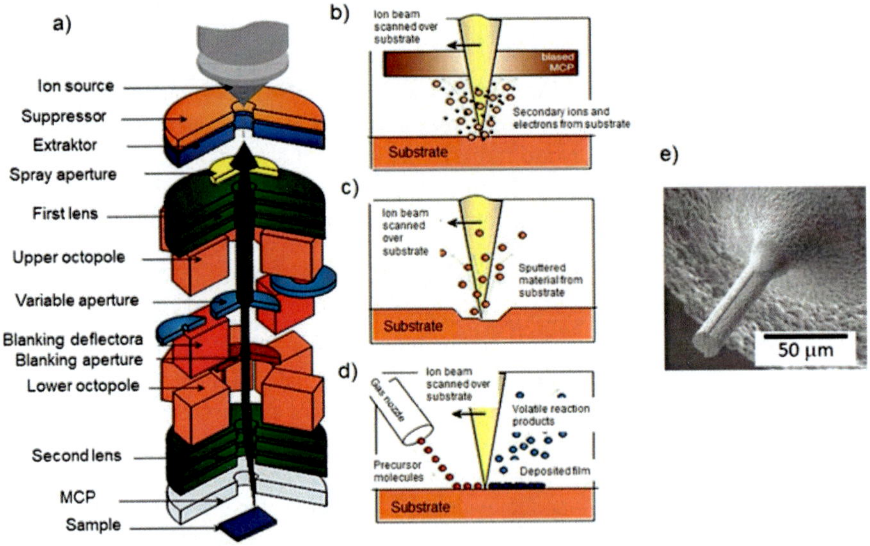

Fig. 4.18 Scheme of equipment for focused ion beam machining (FIBM) **a** and different variants of the process applications: **b** emission of additional ions and electrons from the workpiece; **c** material removal, **d** material deposition—additive shaping, **e** representative example of application [18, 19]

allows scanning the surface (Fig. 4.18b, c, d) with a focused ion beam [10, 18, 19]. The micro-milling semi-finished product shown in Fig. 4.18e, initially fabricated by wire electrodischarge machining (WEDM), was subjected to FIB (Focused Ion Beam) spraying to shape the blades (sample process parameters: accelerating voltage—60 kV, powder grain size 110 μm, ion beam velocity 2200 mm/s; beam current 19 mA) [19].

The electron or ion beam machining processes are widely used in the fabrication of microelements in the electronics, automotive, aerospace, or space industries. Their application in hybrid processes is limited, for example, ion etching assisted by chemical reactions is mainly used [][19].

4.2 Characterization of Mechanical Influence on the Workpiece Material

In subtractive cutting and abrasive machining processes, the mechanical loads occurring in the cutting zone (Figs. 4.1 and 4.4) are usually characterized by the componential cutting forces, specific cutting pressure/specific cutting force, and specific cutting energy/power [1, 2]. In order to evaluate the energy intensity of various subtractive machining processes, the specific cutting energy/power is mainly used, which determines the energy/power consumption per unit volume of material removed (J/mm^3 or GJ/m^3). According to various literature sources [1, 2], specific cutting force takes the following values:

- in the cutting operations of the most common construction materials, the value of $k_c = 1000$ - 5000 N/mm^2 (MPa). Its value is determined by the intensity of plastic deformation in the process of chip formation, the tensile strength/ separation strength of the material, and to a lesser extent by friction and plastic deformation of the surface layer material. It is worth mentioning that the value of k_c is strongly dependent on the thickness of the cut layer which decreases according to a hyperbolic distribution along with its increase.
- in grinding operations, the value of k_c is much higher, which is due to the large contribution of phenomena such as ploughing, crushing and friction (Fig. 4.4). It is estimated that its value can reach 30,000–100,000 MPa in grinding of planes and cylindrical surfaces and 5000–20,000 MPa in cutting-off by wheels. Other sources [2, 4] put the value of k_c for grinding C45 steel in the range of 15,000–60,000 MPa.
- in abrasive polishing, when moving into the nano-machining area, the k_c value can increase even above 10^5 MPa. In this case, it is necessary to use abrasive-assisted electrochemical machining *(AAEM)* or mechano-electrochemical machining (MECM).

The comparison of k_c values for different machining modes of cutting and abrasive machining processes is shown in Fig. 4.19. On the other hand, the value of specific

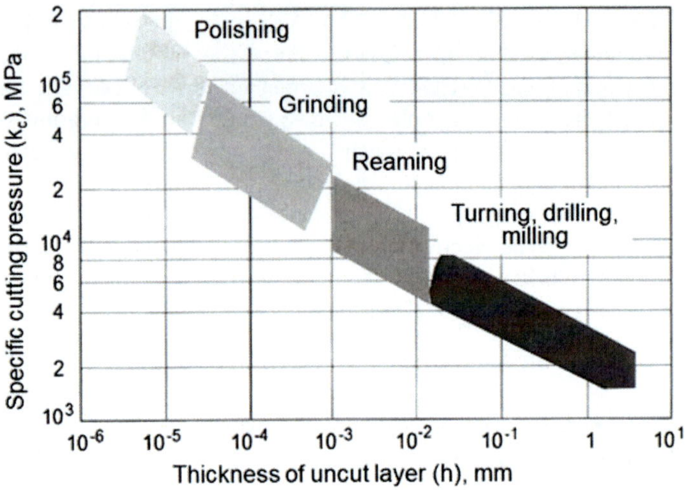

Fig. 4.19 Values of specific cutting pressure for different cutting and abrasive processes [1, 2]

cutting energy depends on the machining process used and thickness of the cut layer as shown in Fig. 4.20a.

According to various literature sources [1, 2], the specific cutting energy/power (e_c/p_c) takes the following values:

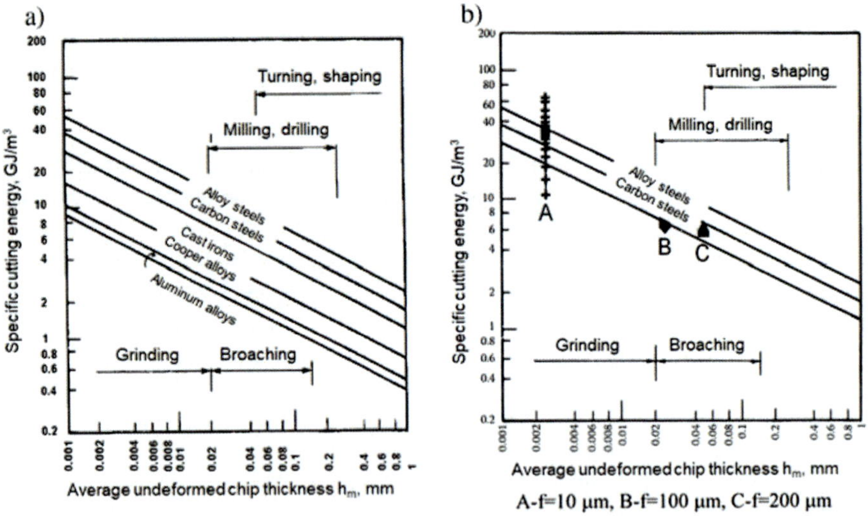

Fig. 4.20 Values of specific cutting energy for different machining operations and machinability groups **a** [1] and the influence of feed rate in the turning of 16MnCrS5 (AISI 5151) steel hardened to 850-800HV$_{0,05}$ **b** [24]

- in cutting process, the e_c reaches values in the range of 0.4–10 J/mm^3 (GJ/m^3). Higher e_c values are characteristic of the machining of difficult-to-machine materials (in particular, high-temperature creep resistant and heat-resistant aerospace alloys of the HRSA group and steels hardened above 50 HRC), which primarily require the additional use of various forms of mechanical, thermal and tribological assistance.
- in surface grinding $e_c = 30–100$ J/mm^3, and in wheel cutting-off $e_c = 5–20$ J/mm^3. It is worth noting that these values are much higher than the value of specific energy of melting the workpiece material, which for steel is equal to $e_m = 13.5$ J/mm^3 [1]. Due to the very high energy consumption of finishing grinding, it is recommended, if technologically possible, to use finishing turning or milling operations, the so-called hard machining [1]. A comparison of the specific energy values for precision hard turning with variable feed rate (thickness of the cut layer) and depth of cut and grinding is shown in Fig. 4.20b. Possible measures to reduce the mechanical loads include reducing the strain hardening of the material by laser or plasma heating, changing the contact conditions in the tool-material zone by applying ultrasonic vibration, and using friction-reducing media such as MQL. The cited data confirm the purposefulness of hybridization of mechanical machining processes also at the micro- and nano-scale, where a lot of energy is lost to overcome the resistances not directly related to overcoming the material cohesion and removing, i.e. ploughing, sliding friction and internal friction.

4.3 Characterization of Thermal Influence on the Workpiece Material

In subtractive cutting and abrasive machining operations, the thermal loads occurring in the cutting zone are usually characterized by the cutting/grinding temperature and the intensity of the heat generated in the form of componential heat fluxes related to the different zones in the chip formation area (Figs. 4.2 and 4.4) [1, 2]. Figure 4.21

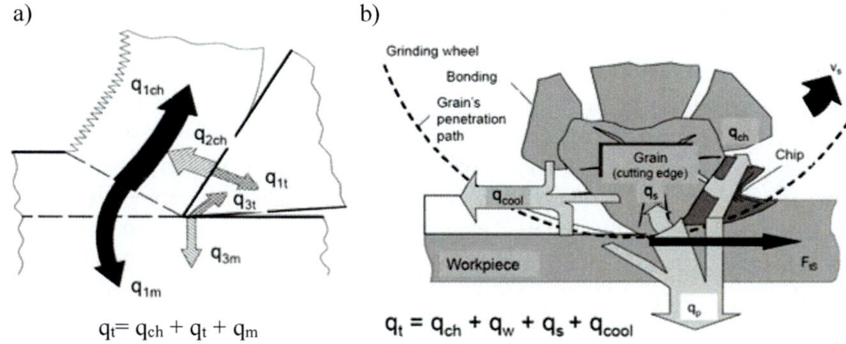

Fig. 4.21 Heat partition in the cutting **a** and grinding **b** zones [1, 4]

shows the heat distribution for cutting and grinding processes.

It is known from well documented experimental research that the heat fluxes q_{ch}, q_t and q_m, penetrating into the chip, tool, and workpiece material, respectively (Fig. 4.21a), and thus the temperature values in the analysed areas of machining zone, depend on the conductivity and heat capacity of the workpiece and tool materials, the cooling method, and the cutting parameters, with the decisive influence of the cutting speed. In the case of a cutting edge with a defined stereometry, the basic sources of heat (energy dissipation) are plastic deformation in the sliping region (1), friction at the contact of the cutting tool with the chip (2), and the friction at the contact of the cutting tool with the workpiece surface (3) [1]. In the case of grinding represented by a single grain (Fig. 4.21b), part of the total heat flux (q_t) flows to the grinding wheel (q_s), the workpiece (q_w), the chips (q_{ch}), and to the coolant (q_{cool}). The distribution of these fluxes depends on the conductivity of the workpiece, the grinding wheel, the cooling fluid, and the heat transfer coefficient [2, 4]. In grinding, both the thermal interaction time and the temperature value at the workpiece surface are satisfactorily controlled by the intensity of the cooling lubricants. The thermal balance of the process can be improved by using various forms of process assistance and combinations. For example, media that reduce heat generation and reduce friction are highly effective, e.g. cryogenic cooling and MQL. In industry, there has been an increase in the use of waterjet machining, which is a "cold" process, as a primary or secondary process (Sect. 4.1.5).

The effects of intense heat dissipation in EDM and laser machining on changes in material properties under the machined surface are presented in Sect. 4.4.

4.4 Constitution of Subsurface Layer of the Workpiece Material

The subtractive machining process creates a new surface characterized by topography, microstructure (metallurgical properties) and mechanical properties [1, 4]. The actual machined surface is very complex and consists of a system of interrelated features that affect its functional properties (usable technological quality), i.e., functionality. In order to determine the different mechanisms of surface formation in subtractive machining, it is proposed to assign them to three individual mechanisms: chemical, mechanical and thermal, or more appropriately to classify them into five groups: chemical, mechanical, mechano-thermal, thermo-mechanical and thermal [1, 26], as shown in Fig. 4.22. It should be kept in mind that the primary mechanisms are always present, to a greater or lesser extent, but with different energetic partition. It can be seen in Fig. 4.22 that the final energy balance covering various energy inputs suggests up to a sevenfold increase in energy entering the surface of material during the process.

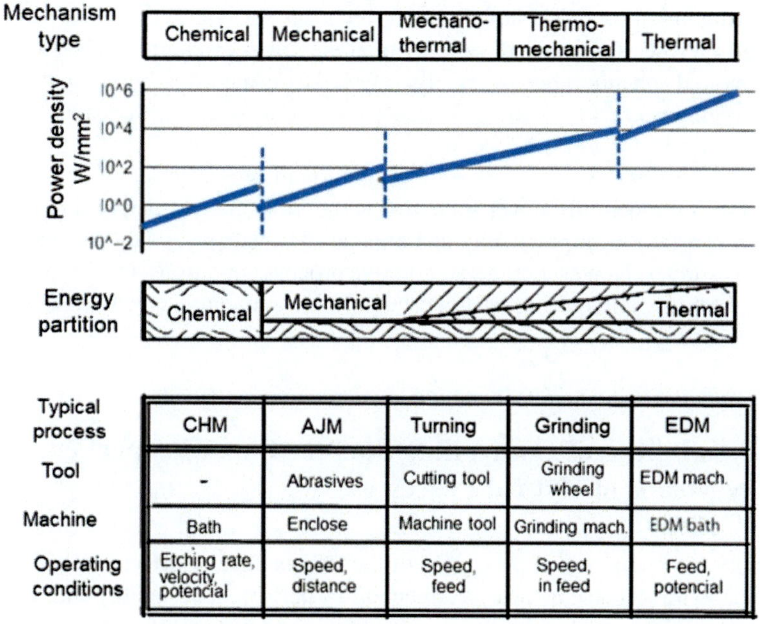

Fig. 4.22 Mechanisms of surface generation in typical material removal processes [1, 20, 21]

Typically, high energy input increases the probability of microstructure defects, which then translate into deterioration of the constituted surface layer. In particular, the surface layer subjected to mechanical impact experiences such changes as sticking, overlapping, folding and plastic deformation. The heat-affected zone (HAZ) exhibits changes such as phase transformations, cracks, and de-hardening, and also the *chemical-affected* zone (CHAZ) is formed by chemical changes in the material on the surface. Furthermore, the stress-affected zone (SAZ) is created as a result of the induced machining stresses that are, in the general case, the result of a combination of mechanical and thermal influences. Considering the level of surface power/energy density (Fig. 4.21), five basic processes can be distinguished: electrochemical machining (ECM), abrasive jet machining (AJM), turning (T), grinding (G), and electrodischarge machining (EDM), at energy levels appropriate to perform each generation mechanism. Additionally, to account for the effects of machining conditions, including cutting speed, feed rate, depth of cut, tool cutting edge condition, and lubrication/cooling conditions, etc., on the surface condition, machining passes can be categorized as aggressive, normal, and gentle. In general, aggressive processes cause deterioration of surface layer quality due to generation of more heat and plastic deformation with higher intensity and speed (examples are EDM and plunge grinding). On the contrary, gentle process conditions result in less heat generation, thus producing surface layer with little or no stress (examples include electrochemical machining or electrochemical assistance of mechanical process, as well as water jet machining).

The structure of the EDM surface is referred to as isotropic because it is formed by spherical micro-craters created by individual electrical discharges. The subsurface layer is melted and amorphous—usually referred to as the *white layer*—and contains numerous micro cracks. Below a heat affected zone with high tensile stresses is produced. Similar thermal effects occur in laser machining (Sect. 4.1.6). In electrochemical process (Sect. 4.1.4), on the other hand, the material is removed atom by atom at a temperature lower than 100 K, so both microstructure and hardness changes in surface layer practically do not occur. This can be assumed to be a *stress-free surface layer*. However, under aggressive process conditions, the electrolyte can weaken the material cohesion at grain boundaries and cause selective etching and corrosion pitting [20, 22].

4.5 Possibilities of Controlling Surface Layer Properties by Means of Hybrid Processes

The geometric and physical properties of the surface layer (surface integrity) are of fundamental importance in the assessment of the performance, quality and reliability of parts shaped by machining processes during their service life. In the context of the assessment of the influence of hybridization on the properties of surface layer it is better to consider individual parameters summarized in three groups, as shown in Fig. 4.23. These are the groups of mechanical, metallurgical and topological properties. In a practical approach, the surface roughness from the group of topological

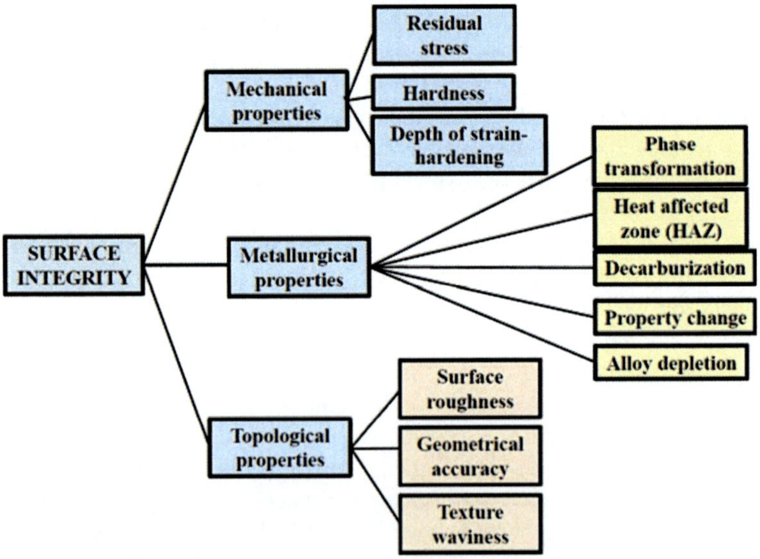

Fig. 4.23 Characteristics of subsurface layer in terms of groups of properties and their parameters

properties and the residual stress and hardness (or more precisely their distribution under the surface) from the group of mechanical properties are of the greatest importance.

The effects of hybridization on the surface layer condition and subsequent functionality are very complex and are usually referred to three basic methods (Fig. 1.6), i.e., assisting with additional energy sources, combining different processes, and controlling the physical mechanisms of the processes. These will be discussed in detail in Chaps. 6–12. In general, hybridization as an effective way to improve machinability (workability) and surface layer quality concerns the following influences [1, 3, 9, 23]:

- changes in the material properties of the machined layer and/or cutting tool during the process. For example, laser heating in LAM machining results in softening of the workpiece material and reduction of plastic deformation in machining aerospace alloys based on titanium, nickel and cobalt, or a transition from a brittle to a ductile state in machining ceramics. The opposite effect is obtained in cryogenic treatment, when the material is hardened by cooling with liquid nitrogen. In both cases, the thermal influence is reduced resulting in a reduction of the heat affected zone and a change in stress distribution.
- changes in the contact conditions between the tool and the workpiece by applying vibrations, electric and magnetic fields, supplying fluids under high pressure or special media improving tribological conditions, e.g. fluids with lubricating additives, e.g. graphene nanoparticles to the oil mist (MQCL). The main positive result can be improved access of the cooling lubricant to the contact zone (USAM, HPC), reduction of friction (dissipation of additional energy) and thus reduction of process temperature. These effects have important implications for subsurface layer properties in all three groups.
- controlling the process so that the required functional properties can be achieved without additional operations. A good example would be grind hardening (Chap. 11), where the heat generated during grinding process is used for final heat treatment of the machined material. A second representative example can be the case of deep rolling with subzero cooling of the surface, which leads to controlled transformation of the microstructure and generation of high compressive stresses in the subsurface layer to increase the fatigue strength of the part. This group of interactions mainly concerns controlling mechanical properties.

References

1. Grzesik W (2017) Advanced machining processes of metallic materials. Elsevier, Amsterdam
2. Toenshoff HK, Denkena B (2013) Basic of cutting and abrasive processes. Springer, Berlin
3. Cheng K, Hou D (2013) Micro-cutting. Fundamentals and applications. Wiley, London
4. Klocke F (2011) Manufacturing processes 2. Grinding, lapping, polishing. Springer, Berlin
5. Brinksmeier E, Riemer O, Gessenharter A (2006) Finishing of structured surfaces by abrasive polishing. Prec Eng 30(3):325–336

6. Barylski A (2013) New disc tools for machining plain surfaces on lapping machines. Mach Eng (Inżynieria Maszyn) 18(2):84–96
7. Kozak J (2018) Mathematical modelling of advanced manufacturing processes. Science Library of the Institute of Aviation no 56 Warsaw
8. Ruszaj A, Grzesik W (2012) Manufacturing of sculptured surfaces using EDM and ECM processes. In: Davim JP (ed) Machining of complex sculptured surfaces. Springer, London, pp 229–251
9. Lauwers B, Klocke F, Klink A, Tekkaya AE, Neugebauer R, McIntosh D (2014) Hybrid processes in manufacturing. Ann CIRP Manuf Technol 63(2):561–583
10. El-Hofy H (2005) Advanced machining processes. McGraw Hill, New York
11. Chavosi SZ, Luo X (2015) Hybrid micro-machining processes: a review. Prec Eng 41:1–23
12. Luo X, Qin Y (2018) Hybrid machining. Theory, methods and case studies. Elsevier Academic Press, London
13. Internet: Producing microstructures with a laser/scanner system. Industrial laser solutions. www.google.com/search
14. Davim JP (ed) (2012) Machining of complex sculptured surfaces. Springer, London
15. Kalpakjan S, Schmid S (2013) Manufacturing engineering and technology. Pearson, Chicago
16. Internet: electron beam machining—HIDAYAT (KQG 170006)
17. Anwar S, Ahmed N, Abdo BM, Pervaiz S, Chowdhury MA, Alahmari AM (2018) Electron beam melting of gamma titanium aluminide and investigating the effect of EBM layer orientation on milling performance. Int J Adv Manuf Technol 96:3093–3107
18. Internet: https://www.bing.com/images
19. Ali MY (2008) Focused ion beam micromachining of MEMS. International symposium on advances mechanics engineering February 2008. In: CNU Korea Changwon
20. Grzesik W, Ruszaj A, Kruszyński B (2010) Surface integrity of machined surfaces, Chapter 5. In: Davim JP (ed) Surface integrity in machining. Springer, London
21. Griffiths B (2001) Manufacturing surface technology. Penton Press, London
22. Burakowski T, Wierzchoń T (1999) Surface engineering of metals. Principles, equipment, technologies. CRC Press, London
23. Grzesik W (2016) Prediction of the functional performance of machined components based on surface topography: state of the art. J Mater Eng Perf 25(10):4460–4468
24. Grzesik W, Denkena B, Żak K, Grove T, Bergman B (2016) Energy consumption characterization in precision hard machining using CBN cutting tools. Int J Adv Manuf Technol 85:2839–2845

Chapter 5
Modelling of Hybrid Machining Processes

5.1 Models of Conventional and Unconventional Machining Processes

5.1.1 Models of Cutting and Abrasive Processes

Material removal in cutting and abrasive processes takes place under complex thermo-mechanical, and often thermo-mechanical-chemical conditions (Sect. 4.1.1), which determines the appropriate consumption of mechanical energy to overcome material cohesion and friction [1, 2]. It should be noted that the physical phenomena occurring in the machining zone are, on the one hand, closely interrelated due to mutual interactions and, on the other hand, are distinguished by a particularly high intensity, as illustrated in Fig. 5.1.

It is generally accepted that material in the cutting zone (primary and secondary plastic deformation zones) is subjected to extreme values of strain and strain rate, about $\varepsilon = 8$ and $\dot{\varepsilon} = 10^5\ \mathrm{s}^{-1}$ respectively, together with temperatures exceeding 1000° and temperature gradients of the order of 2×10^6 K/s. Such extreme contact conditions in the tool-chip interface cause normal stresses to reach values on the order of 3.5 GPa [1–3]. It should be noted that in grinding the strain rates, due to grinding speeds reaching 200 m/s, are much higher, i.e. in the range of 10^5–10^7 s^{-1}. Correspondingly, during grinding the temperature gradients in the subsurface layers reach values of about $10^{6}\ {}^{0}$C/s [1, 5].

In grinding (Fig. 5.2), the problem of modeling the individual cases (I–IV) of chip formation by micro-cutting is more complex than in orthogonal cutting, also due to the need to select a representative stereometry of a single abrasive grain. Therefore, as shown in Fig. 5.2, the grinding speed and the depth of material removed depend on the shape of the abrasive grain. Typically, the chip formation model assumes a grain with a conical cutting edge, less commonly a spherical one, and the slip-line field analysis or numerical methods [5] are used. These issues were previously described in Sect. 4.1.2.

W. Grzesik and A. Ruszaj, *Hybrid Manufacturing Processes*, Springer Series in Advanced Manufacturing, https://doi.org/10.1007/978-3-030-77107-2_5

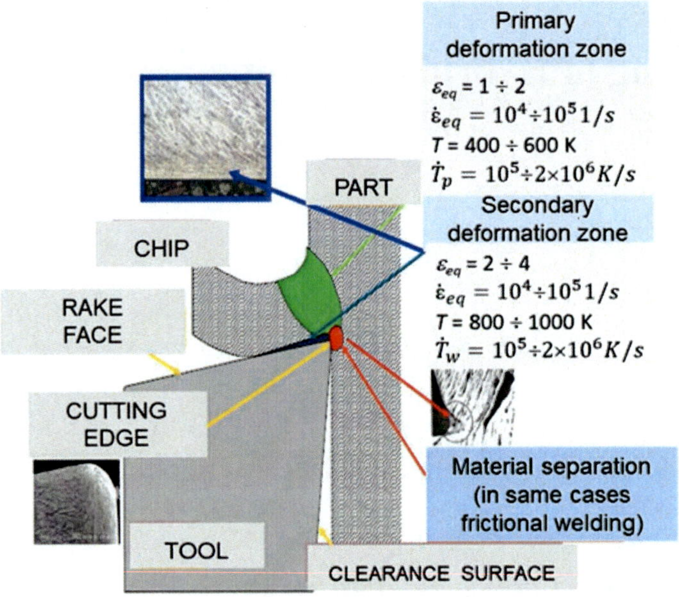

Fig. 5.1 Schematic illustration of thermo-mechanical conditions in the cutting zone with developed (fan-type) shearing area [3, 4]

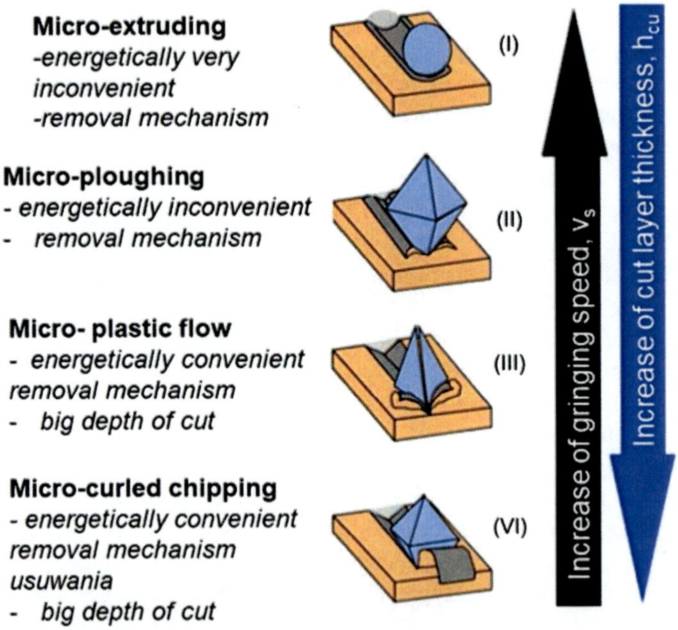

Fig. 5.2 Schematic illustration of material separation conditions with single abrasive grain in the grinding zone [5]

The presented peculiarities of the chip formation process in machining and grinding require for their experimental identification very advanced techniques and devices, e.g. the *Split Hopkinson's Pressure Bar* technique (SHPB device), Taylor's impact test or special tribo-testers [1, 3]. In hybrid processing, there are also problems of modeling additional sources of vibration energy, thermal energy due to laser or plasma heating, heat extraction due to cryogenic media cooling, and hydrodynamics of fluid flow under high pressure [1]. As a result, modeling the coupled mechanical-thermal state of a material with acceptable accuracy is still a major challenge, and increasingly complex, engineering task due to the presence of many physical and technical constraints.

5.1.2 Models of EDM Process

In electrodischarge machining (EDM), material is removed during electrical discharge as a result of current flow in the plasma channel, heat release, temperature rise and thus by melting and partially evaporating the electrode material and dielectric. A scheme of such a discharge area with a plasma channel of radius r_p is shown in Fig. 5.3. The expansion of the plasma channel in time is described by the following empirical relation [6]:

$$r_p = 0.788 t^{\frac{3}{4}} \tag{5.1a}$$

An indicator of EDM machinability and the suitability of materials for electrode tools can be their EDM resistance *S*, which has been defined in works [7], among others:

$$S = (1 - \nu)\frac{T\lambda}{E\alpha} \tag{5.1b}$$

where: *S*—material electrodischarge resistance, *E*—Young's modulus, *T*—melting point, α—linear expansion coefficient, λ—thermal conductivity coefficient, ν—Poisson's ratio.

The electrode material should be characterized by as high an electrodischarge resistance as possible. For the selected electrode material, the electrode wear depends significantly on the process parameters, especially on the characteristics of the electric pulses and on the shape and dimensions of the workpiece surface. The shape and dimensions of the electrode should be adjusted considering the width of the inter-electrode gap and the electrode wear. Studies of the influence of electrode material properties on the erosion process have shown that in most cases it is best to use graphite electrodes for roughing and copper electrodes for finishing. Of course, in some special cases, other materials should be used, for example, tungsten or composite materials produced by metal powder sintering [7]. In the EDM sinking process, the electrode is displaced in the workpiece direction so that stable electrical discharges occur between the workpiece and the electrode tool. Usually, the capability of the driving system is such that it can maintain a constant gap in the range of 100–500 μm. The gap width in the EDM process results from the parameters of the electric pulses, and these are selected depending on the material removal rate and accuracy requirements and so as to eliminate arcing, short-circuit or empty pulses [7, 8].

The EDM machining conditions are primarily determined by the value of the voltage pulse amplitude (usually 60–120 V), the amplitude of the operating current pulse (usually <1–120 A), the voltage pulse time and interval time (usually <1–2000 μs), the type and physical properties of the dielectric and its flow conditions (dielectrics based on petroleum products, deionized water), the type and physical properties of the workpiece material and the electrode tool [6–12]. The EDM sinking process is usually carried out in a "tank–bath" filled with liquid dielectric. The dielectric continuously flows through the bath and partially through the machining zone, removing erosion products and stabilizing the machining conditions. The EDM process can be used, among others, for machining working cavities of dies, molds or injection molds and other shaped surfaces on macro, meso and micro scale [5–12].

The energy required for the EDM process is obtained from an electrical pulse generator. The higher discharge energy results in a larger volume of material removed and a higher efficiency, but at the same time the surface roughness deteriorates. However, regardless of the pulse characteristics, the amount of material removed during a single discharge is very small. Therefore, in order to achieve satisfactory performance, the pulse frequency and thus the discharge frequency is selected in the range of 10^3–10^9 Hz. Modern control systems provide electrode consumption as low as less than 1% and material removal rate from about 0.1–400 mm^3/min. The first applications of EDM included primarily sinking operations, and now also cutting and milling operations are applied [6–11].

It follows from the above characteristics that in order to properly design and implement the EDM process, it is necessary to have accurate technological characteristics. At the present stage of EDM development such characteristics are usually determined by machine tool manufacturers based on experimental-mathematical statistical models. Of course, since the beginning, theoretical studies of the EDM process have been carried out to develop a mathematical-physical model of the phenomena occurring during a single discharge and multiple discharges as they occur in real machining conditions. Theoretical studies use the fundamentals and achievements of electro-mechanical, thermo-mechanical and thermo-electrical theory to explain and describe the EDM erosion process.

Due to the fact that many complex phenomena occur in the EDM process (often simultaneously), building a general physical model of the EDM process has not yet been possible. In the theoretical studies carried out, one usually starts with the study of the phenomena occurring in individual discharges (Fig. 5.3), and the examination begins with the determination of the energy distribution between the workpiece material, the electrode tool and the dielectric. Later, the temperature distribution (Fig. 5.4), the mass of material removed from the workpiece and the electrode tool, the hydrodynamic conditions in the discharge region, and the crater geometry are

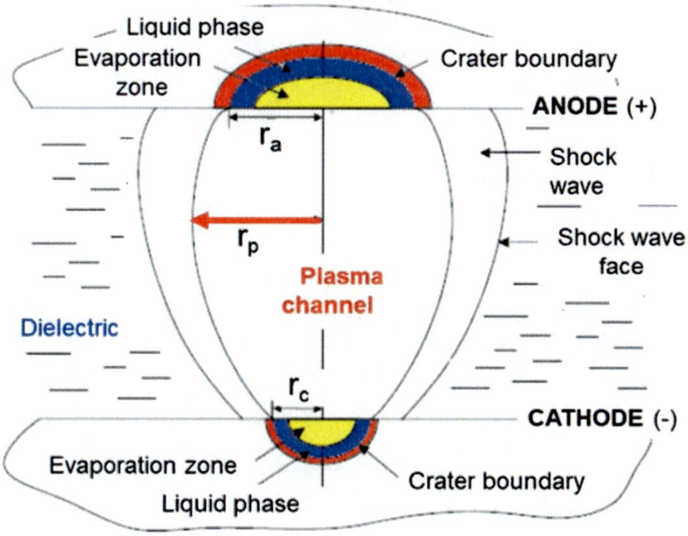

Fig. 5.3 Scheme of discharge (plasma) channel [7, 8]

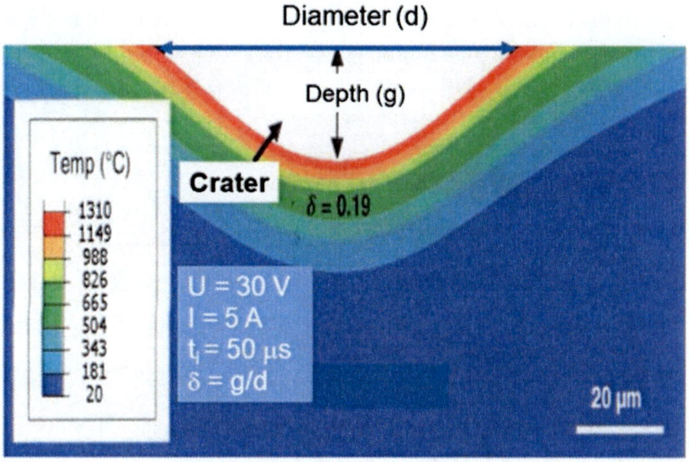

Fig. 5.4 Crater geometry and temperature distribution after single discharge obtained by means of theoretical (mathematical) model [7]

determined. This allows an attempt at determining the distribution of subsequent discharges and then, for a known discharge frequency, predicting the efficiency and wear of the electrode.

The results of theoretical studies are still too inaccurate to design the technological process in industrial conditions on their basis. However, it is clear that there is a feedback between the mathematical modeling of the phenomena occurring in the

EDM process and the experimental determination of technological characteristics. This is an important part of EDM technology development. Analogous research is carried out in sinking, milling or wire-cutting processes. In milling or wire-cutting processes, the temporary machining area is much smaller than in sinking because the area of the electrode tool is reduced, and the shape of the machined surface is a reproduction of the electrode trajectory, not its shape and dimensions as in the sinking operation. From the above discussion, the limitations of EDM are related to the variety and complexity of physical and chemical phenomena accompanying the material removal, wear of the electrode tool, and process control. These limitations can be mitigated by hybridization, i.e., supporting the EDM process with electrode vibration (EDM–USM), electrochemical machining (ECDM), abrasive machining (AEDM) or laser machining (LEDM). These processes will be described in Chaps. 6, 8, 9 and 10.

5.1.3 *Models of ECM Process*

Most often, electrochemical contactless machining is used in the sinking operation (Fig. 5.5), in which the electrode tool moves toward the workpiece at a constant speed, usually in the range $v_f \cong 0.1 - 1.0$ mm/min. Through a hole in the electrode, electrolyte is injected into the machining area at pressures as high as 3 MPa. Pressure and electrolyte flow are adjusted in such a way as to remove from the machining

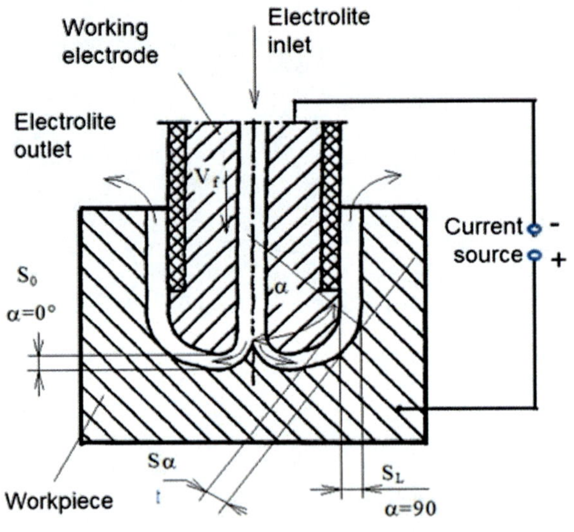

Fig. 5.5 Scheme of ECM sinking of curvilinear surface, i.e. working mould or die cavities when side walls are isolated in order to side workpiece surface against electrochemical dissolution; α—profile angle

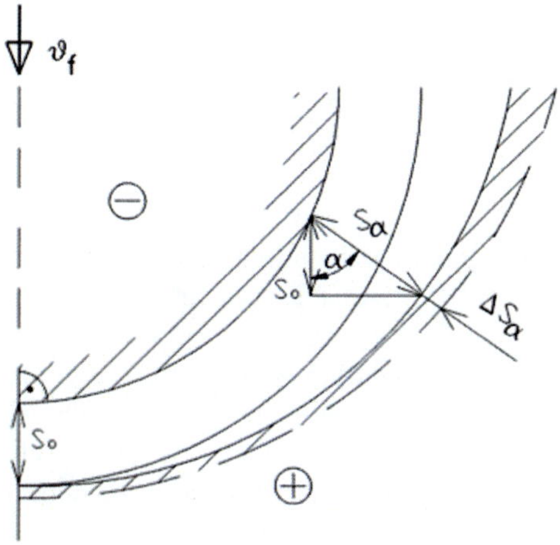

Fig. 5.6 Geometry of inter-electrode space in a stable state; "–"—cathode, "+"—anode, v_f—velocity of electrode movement, α—profile angle, ΔS_α—change of gap thickness resulting from process conditions in the selected cross-section, S_0—gap thickness for the angle $\alpha = 0$

area the heat generated by electric current flow, hydrogen released on the cathode surface and products of anodic reactions. The process can be used, among others, for machining working cavities of dies, molds and injection molds. The area of the machined surface in this case may be from several to even several hundred cm^2. To obtain a satisfactory quality of the machined surface the current density should be several dozen A/cm^2. This results in the fact that the amperage during machining may be several hundred and in extreme cases even several thousand amperes. Under such conditions, the basic problem is to ensure appropriate hydrodynamic conditions in the inter-electrode gap, the width of which, depending on the profile angle α and the local electrical conductivity of the electrolyte, may vary from about 0.1 mm (for $\alpha = 0°$) to several tenths for larger values of the angle α [7, 10, 11, 13].

At any given point (Fig. 5.6), the rate of change in gap width *ds/dt* is equal to the difference in the rate of electrode advancement in the direction normal to the work surface and the rate of material removal by anodic reactions, as noted by Eq. (5.2).

$$\frac{\mathrm{dS}}{\mathrm{dt}} = \eta \mathrm{k}_v \kappa \frac{(\mathrm{U} - E)}{\mathrm{S}} - \mathrm{v_f} \cos\ \alpha \tag{5.2}$$

where: S = S(t)—gap width at the time t, U—inter-electrode voltage, E—sum of potential drops in the interelectrode layers, v_f—electrode feed rate, η—current efficiency of the dissolution process, k_v—volumetric electrochemical equivalent of the workpiece material–anode, α—profile angle.

After a certain time the ECM process reaches a steady state, which means that the geometry of the machining space (Fig. 3.1) does not change and the process indicators do not depend on time, i.e. $ds/dt = 0$. Then the solution of Eq. (5.2) is correlation (5.3a), or correlation (5.3b) if a stationary electrode is used.

$$S_u = \frac{\eta k_v \kappa (U - E)}{v_f \cos \alpha}, \tag{5.3a}$$

$$S = \sqrt{2\eta k_v \kappa (U - E)t + S_0^2} \tag{5.3b}$$

where: S_u—steady-state gap thickness, $S = S(t)$—gap thickness at the time "t", S_0—inter-electrode gap thickness at the initial time of the process $t = 0$.

The above Eqs. (5.3a) and (5.3b) enable to determine the distribution of the inter-electrode gap width and thus to carry out the correction of the shape and dimensions of the electrode tool. To determine the actual distribution of the inter-electrode gap width, the distributions of: electrolyte flow velocity and pressure, hydrogen volume concentration C_H, electrolyte temperature T and its specific electrical conductivity must be determined. Optimally selected hydrodynamic conditions prevent the formation of cavitation zones, circulation, stagnation, passivation and excessive increase in temperature (ΔT) and hydrogen volume concentration (CH), which reduces the probability of electrical discharges leading to electrode damage and process interruption [7, 13].

Mathematical modeling of the ECM sinking process of 3D surfaces is a very difficult issue. Of course, mathematical models of the ECM sinking process for the general case have been developed [7, 12, 13], but their practical utility is limited. This is due to the fact that the ECM sinking process is usually used in batch production, but as already emphasized, the electrode tool at optimal and stable process parameters does not wear. That is, once designed electrode is used to make hundreds or even thousands of components, such as turbine blades of aircraft flow engines.

In the case of small batch production or high accuracy requirements, it is necessary to look for such variants of machining, in which the influence of hydrodynamic conditions on the inter-electrode gap thickness distribution is insignificant. One of the possible rational solutions is the application of pulse voltage in electrochemical machining [7, 12, 13]. In the interval between voltage pulses, heat and electrode reaction products are extracted so that the next voltage pulse is generated in a clean electrolyte. The effect of hydrodynamic conditions on the ECM process decreases as the machined area is reduced, i.e., when starting micro-element sinking. Another option to reduce the machining area is to reduce the electrode dimensions and shape the workpieces in planing or electrochemical milling operations with a flat cylindrical or spherical electrode with a small working area [7, 12–14] or wire cutting [15]. In planing, milling or cutting operations, the shape and dimensions of the workpiece are a reproduction of the trajectory of the electrode tool.

From the above discussion, it can be seen that ECM machining primarily has limitations related to the design of hydrodynamic conditions and the removal of heat and electrode reaction products from the machining area. These limitations can be mitigated by hybridization, i.e., supporting the ECM process with electrode vibration (ECM-USM), electro-discharge machining (ECDM), abrasive machining (AECM), laser machining (LECM) or high-pressure electrolyte and abrasive jet (AWJECM).

5.1.4 Models of LBM Process

Laser beam machining (LBM) is a widely used thermal method for subtractive machining, additive machining, forming or heat treatment. In subtractive machining, it is used in operations such as drilling, cutting, grooving, turning and milling (Fig. 5.7). Most commonly, CO_2 molecular laser and pulsed solid-state Nd:YAG laser are used as heat source, but also excimer laser as well as pico and femtosecond lasers are also used in micromachining [1, 16, 17]. In the mentioned laser machining operations, different material removal mechanisms are used. It should be noted that drilling should be considered as a one-dimensional problem, while cutting, turning and milling as a three-dimensional problem. It requires the simultaneous use of two laser beams (*A* and *B* in Fig. 5.7), each of which makes a blind groove around the perimeter (b1 and b2) or along the surface of the workpiece (c), and material is removed in the form of a wedge by their intersection. This arrangement results in a significant increase in material volume per unit energy compared to single laser beam machining. Material removal in LBM processing involves three stages (Sect. 4.2.6), i.e., melting, evaporation, and chemical degradation, which includes the destruction of chemical bonds and the degradation of the material [16].

The thermal energy required to melt and vaporize the material is stored (accumulated) due to absorption of the high-density laser beam, which is focused on the material surface by a focusing lenses system. Additionally, a high-pressure assistive gas can be used to remove molten, vaporized, and chemically degraded material from the machining zone (Fig. 5.8). This results in a clean cutting surface with sharp edges. The formation of the erosion front occurs at the front of the laser beam and for this reason the *thermal field* moves as a result of the relative motion of tool and workpiece. Thus, laser cutting represents a fixed thermal process, however, the temperature of the workpiece is independent of time but is dependent on the distance to the erosion front. The accumulated molten material can be ejected by the assisting gas through the gap or vaporized.

The principle of material removal and thermal energy dissipation is similar in laser drilling and cutting, as well as laser grooving and cutting. Hence, the obvious similarity of laser turning and milling (Fig. 5.7) to laser grooving also follows.

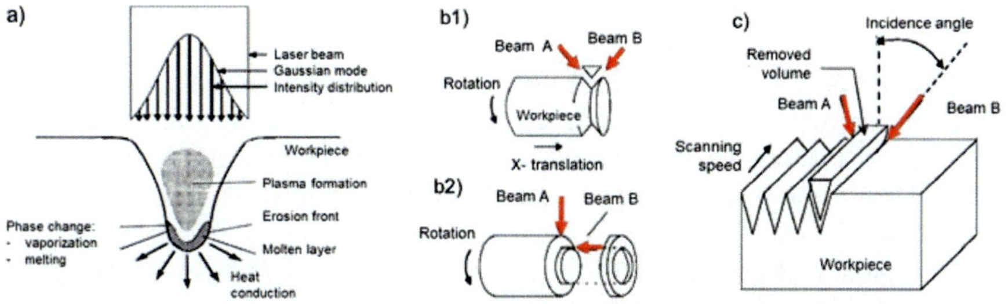

Fig. 5.7 Principles of 1D and 3D laser machining **a** drilling; **b1** helical turning; **b2** ring turning; **c** milling [16]

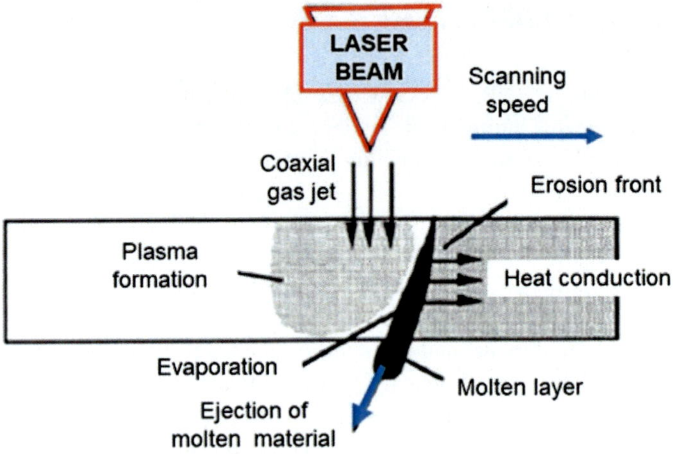

Fig. 5.8 Principle of material removal in laser cutting [16]

Typically, in laser heat flux analysis, boundary conditions are assumed that locate the surface on which the conduction of heat to the workpiece and its loss by radiation, evaporation, and convection take place. The basic conduction equation in operator form adapted to laser cutting has the following form:

$$\rho C u \frac{\partial T}{\partial x} = k\nabla^2 T \tag{5.4a}$$

$$\rho C \frac{\partial T}{\partial t} = k\nabla^2 T + Q \tag{5.4b}$$

where: u—scanning speed of laser beam, T—temperature, t—time, C—specific volumetric heat, x—coordinate of laser beam motion, Q—additional heat loss.

However, for laser drilling, time is introduced instead of the laser beam direction coordinate and then Eq. (5.4a) will change to (5.4b). The boundary condition for the vaporization zone used in Eq. (5.4a) is formulated:

$$I_o \exp\left(-\left(x^2 + y^2\right)/R_o\right) = h(T - T_\infty) - k\nabla T - \rho h_{ig} u \tag{5.4c}$$

The laser beam is assumed to have the shape of a Gaussian bell curve (Fig. 5.7) in Cartesian coordinates x and y, and then its intensity I_o in the focal point (*laser peak intensity*) has the form (5.4c), where R_o is the radius of the concentrated laser beam, h is the natural convection heat rate, h_{ig} is the heat of sublimation, and u defines the convection velocity.

The approximate minimum power required to melt the surface layer of the material is determined from the following correlation [18]:

$$q_{melt} = \frac{2\sqrt{\pi} k r T_{melt}}{A} \quad (5.5)$$

where: A—absorptivity of the workpiece, k—thermal conductivity of the workpiece, r—laser beam radius, T_{melt}—melting point.

As noted in Sect. 4.2.6, lasers are also often used to support a variety of unconventional and conventional machining processes, including pre-heating of the machined layer (thermally assisted machining-TAM). TAM machining processes will be described in Chap. 9.

5.1.5 Models of WJM Process

Comparative analysis of material removal in different processes shows that the ideal tool for shaping workpieces is a jet of liquid with a small cross-section directed under high pressure at the workpiece surface. Material removal takes place at a low temperature (T < 100 °C), so no significant changes in the properties of the surface layer are observed. In modern machine tools, the high-pressure jet of water or water mixed with abrasives has a small cross section and the possibility of precise control of the depth and direction of penetration. It is worth mentioning that among the discussed shaping methods, only electrochemical contactless machining does not introduce significant changes in the surface layer of the processed material, but there are difficulties in localizing the process of dissolution on a small area. For this reason, high-energy liquid jet is considered almost ideal for cutting various materials (Sect. 4.2.5), especially water or its mixture with abrasive grains.

As the pressure of the water abrasive jet in many industrial machines can be as high as 700–1000 MPa, it has a significant kinetic energy which, when it comes into contact with the workpiece, effectively acts on the workpiece surface causing its dynamic erosion. A significant portion of this energy is stored in the abrasive grains and can therefore be used to machine hard materials with high mechanical strength as well as soft materials.

The modeling of the abrasive water jet cutting process is based on the principles of fluid mechanics, in particular Bernoulli's principle. The basic principle of conservation of energy is applied to an ideal (incompressible) fluid as follows:

$$p + {}^{1}\!/_{2}\, (\rho_w v_w)^2 + \rho_w g h = const \quad (5.6)$$

where: p—water pressure, ρ_w water density, v_w—water velocity, g—gravitational acceleration, h—eight of the location above the reference plane. Observing the outflow of water under high pressure, Eq. (5.6) can be used to determine the velocity of the water jet out of the nozzle in air (Fig. 5.9).

Based on the observation of pressurized water outflow into the air and using Eq. (5.6), the jet outflow from the nozzle can be determined for a known water pressure, namely:

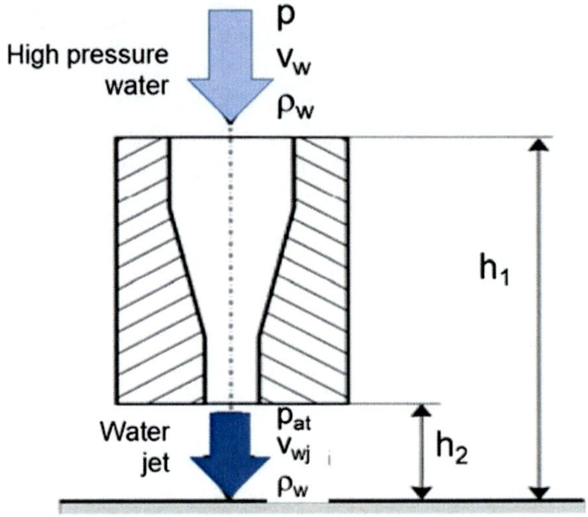

Fig. 5.9 Adaptation of Bernoulli's equation to characterization of WJM process [19]

$$p + 0.5\ \rho_w (v_w)^2 + \rho_w g h_1 = p_{at} + 0.5(\rho_w v_{wj})^2 + \rho_w g h_2 \tag{5.7}$$

After making simplifying assumptions, including that the jet pressure $p \gg p_{at}$ and that $h_1 \sim h_2$, the relationship between jet pressure and jet velocity was obtained, which, along with the nozzle diameter, are fundamental process parameters [19]:

$$p = 0.5\ \rho_w \left(v_{wj}\right)^2 \tag{5.8}$$

Further calculations showed that the computed efficiency is proportional to the power of the water jet (and thus the interaction time) and the nozzle diameter. Mathematical models were also developed to simulate the surface topography—an example is shown in Fig. 5.10b [20].

Mathematical modeling of the AWJM process supports the design of technological processes for cutting materials, however, due to the great variety of materials

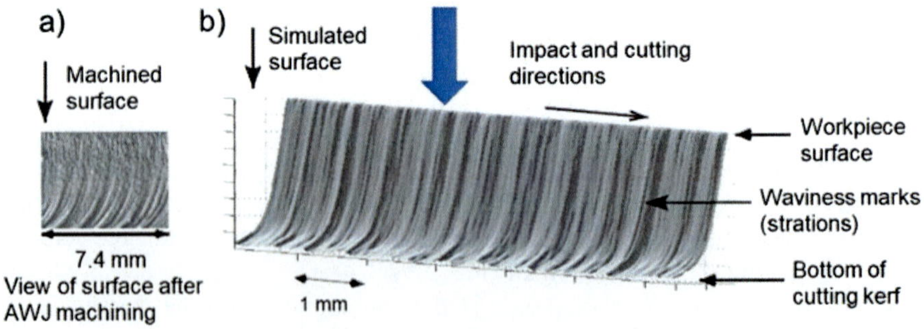

Fig. 5.10 Examples of real **a** and simulated **b** surfaces generated by AWJ cutting [20]

to be machined and the shapes of the manufactured parts, the results of mathematical modeling are not always sufficiently consistent with the results of experimental studies. For this reason, mathematical statistical models developed based on experimental results are also used in the industry. Based on the available mathematical modeling results and experimental results, the cutting process can be described as follows. The jet first penetrates the workpiece material until the full penetration depth of the jet is reached [19–21]. As time progresses due to the jet moving at a constant speed and due to the pressure drop resulting from the jet penetrating deeper in the material, the front of the cutting kerf curves. The angle between the original direction of the jet and the tangent to the cutting kerf face (tilt angle) increases as the jet penetrates into the material. Therefore, the jet deflects more and more in the direction opposite to the direction of the jet feed (Fig. 5.10b). Since abrasive particles have more inertia than water particles they change the direction of flow to a lesser extent. Therefore, the abrasive water flow is locally split and the abrasive particle impacts are concentrated on the part of the slot with the largest tilt angle. In the area where the impact of abrasive particles is concentrated, the material removal rate increases, which is the reason for the formation of a step on the kerf face. This increases the jet deflection and thus leads to a decrease in the material removal rate below the notch. The above description of the high-pressure water jet cutting process has been verified [21].

In a cutting operation, the parameters should be selected to keep the depth of penetration of the jet into the workpiece materialgreater than the thickness of the material being cut. In the case of shaped surfaces, surface waviness can be minimized by an optimal selection of pressure, jet velocity and the size and shape of the nozzle cross-section. Currently available equipment allows obtaining jet pressures up to $p > 1000$ MPa at an abrasive flow rate of 2–8 dm^3/min (abrasive flow rate: 0.4–0.6 kg/min). The jet diameter even up to about 0.025 mm [19–21].

The *AWJM* process has been intensively developed in many research centers, also in Poland. One of the research directions is the assistance of other processes, for example conventional cutting operations (Chap. 7) or electrochemical machining. These issues will be presented in the following chapters.

5.2 Constitutive Materials Models

The success of the process modeling depends on the reliable input data supplied to the simulation program, especially the material constitutive model adapted to the high deformations, strain rates and high temperatures occurring in the chip formation zone (Fig. 5.1). Currently, the most commonly used models include the Johnson–Cook model (J–C model), the Oxley model, the power model used in the commercial program *AdvantEdge*, the J–C model extended by the material fracture module, or even changes in microstructure and hardness [1–3].

The classical J–C model consisting of three terms–elastic–plastic, viscous-plastic and thermal softening is noted in the following form [1]:

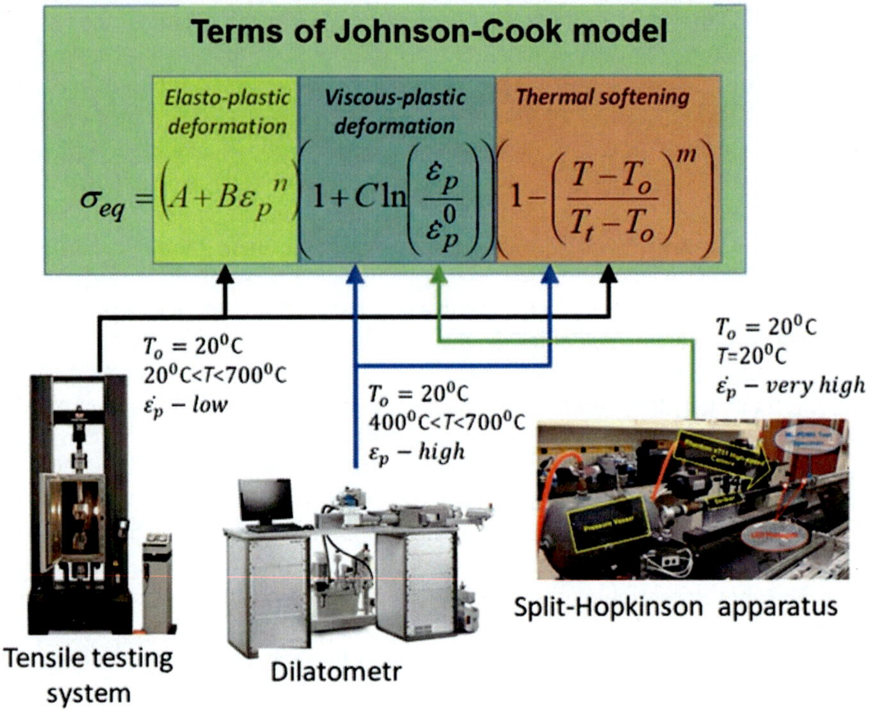

Fig. 5.11 Experimental determination of constants in three terms of the J–C model [22]

$$\sigma_{eq} = \left(A + B\varepsilon_p{}^n\right)\left(1 + Cln\left(\frac{\dot{\varepsilon}_p}{\dot{\varepsilon}_p^0}\right)\right)\left(1 - \left(\frac{T - T_r}{T_m - T_r}\right)^m\right) \tag{5.9}$$

where: ε_p—equivalent plastic strain, $\dot{\varepsilon}_p$—equivalent plastic strain rate,$\dot{\varepsilon}_p^0$—reference strain rate equal to 1.0 s^{-1}, T—temperature, T_r—reference temperature, T_m—melting point of the material, A,B,C- constants determined experimentally for the given workpiece material (the method of their determination is shown in Fig. 5.11).

5.3 Techniques of Determination of Material Properties Under Complex Physical Interactions

Currently, the plastic flow of metallic materials under thermo-mechanical conditions corresponding to chip formation (Fig. 3.1) is experimentally investigated by three methods [1]:

(1) by impact compression tests, such as the Taylor test, recommended when the strain rate exceeds 10^4 s^{-1}.

(2) on a device called a *Split Hopkinson's Pressure Bar-SHPB*, on which strain rates up to 10^5 s^{-1} can now be achieved.
(3) by performing special cutting tests, such as a quick-stopping device during which the tool is rapidly evacuated from the cutting zone and the chip root can be obtained.

Due to hardware limitations, it is advisable to extrapolate the results obtained for lower strain rates, e.g. 10^{-1} - 10^1 s^{-1} even to values of 10^4–10^5 s^{-1} by numerical (FEM method) solution of the inverse problem.

Figure 5.11 shows the original method for determining the constants in the J-C Eq. (5.9) using different material testing techniques and apparatus, i.e., tensile testing system, dilatometer, and SHPB device. It is found that one can combine these techniques effectively and determine the constants *A, B, C* with high accuracy for selected machined materials or obtain data for an empirical model when reliable literature data are not available. Thus, constitutive equations in empirical form were determined for titanium alloy Ti6Al4V and nickel alloy Inconel 718 for machining simulations in aviation industry [22].

5.4 Modelling Techniques

In practice, there is a need to model the process as a structural element of the machining system at different scales, i.e. macro, meso and micro, as shown in Fig. 5.12a. The *macro-scale* model provides data on the distribution of stresses, deformations and temperature in the workpiece-tool system, which is used in practice for *on-line* CAD/CAM control program corrections and process monitoring. The *meso-scale* model is the primary source of information on the course of physical phenomena in the machining zone, such as chip formation or thermo-mechanical conditions in the tool-chip interface. These data are helpful in selecting machining conditions, e.g. to control tool wear. On the other hand, the *micro-* and often even

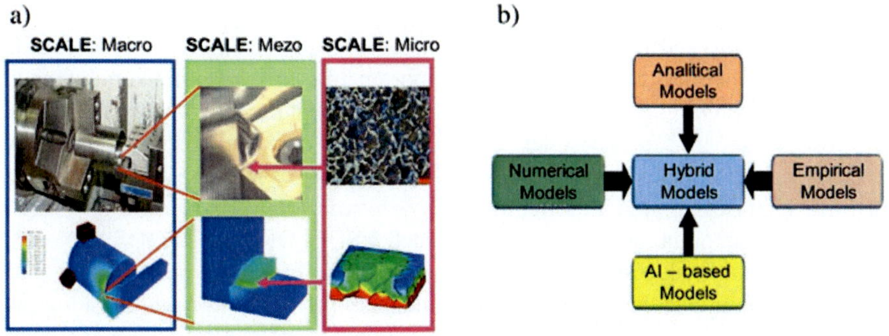

Fig. 5.12 Scale-based division of machining models **a** and classification of models used in investigations of material removal processes **b** [3, 4]

nano-scale model is used to study the changes occurring in the material during the process, such as changes in microstructure, hardness or dislocation motion. Thus, it can be useful for predicting the mechanical properties of the surface layer.

A classification of predictive models, which is commonly used in subtractive machining, is shown in Fig. 5.12b. It contains analytical, empirical, numerical and artificial intelligence (AI)-based models, and by combining them, e.g. analytical and numerical models, hybrid models are created. The different models will be briefly discussed in the following chapters, while the reader will find more information in the books [1, 2].

5.4.1 Analytical Models

Using analytical models (also mechanistic models) based on the identification of physical phenomena occurring in the machining process, it is possible to predict such quantities characterizing the machining process as componential forces, friction, stresses, plastic strain and temperature, among others. An example is the prediction of the cutting force from the value of the specific cutting pressure $k_{c1.1}$ or the dynamic cutting force as a function of the instantaneous value of the cross-sectional area of the cut layer. One of the better known analytical methods is the slip-line field method. However, due to the complexity of machining processes, especially the numerous interactions occurring in the technological machining system [1], they are not useful for predicting all output characteristics, e.g. the change in the accuracy of the workpiece dimensions compared to the semi-finished part, or the quality of the machined surface, which are of more significance in production. It should be added that data from analytical modeling provide valuable data for numerical models (hybrid modeling).

5.4.2 Numerical Modelling

Nowadays, numerical techniques, mainly finite element method (FEM) and finite difference method (FDM), are of great importance in the modeling/simulation of subtractive machining processes. They are very useful in process studies as they are less costly compared to experiments and give results at an engineering acceptable level for studying process mechanics and its thermal behaviour. Typically, FEM methods such as Lagrangian, Euler or ALE (*Arbitrary Lagrangian–Eulerian*) [1, 2] and many commercial software packages such as DEFORM, ABAQUS, ANSYS, AdvantEdge are used. However, an essential condition for reliable simulation results is adequate input data, including the material constitutive equation and the adopting of the finite element mesh to the specific test object (Sect. 5.2). Figure 5.13 shows the results of FEM simulation of the chip formation process in cutting with a chip breaker (a) and single grain grinding (b).

a) b)

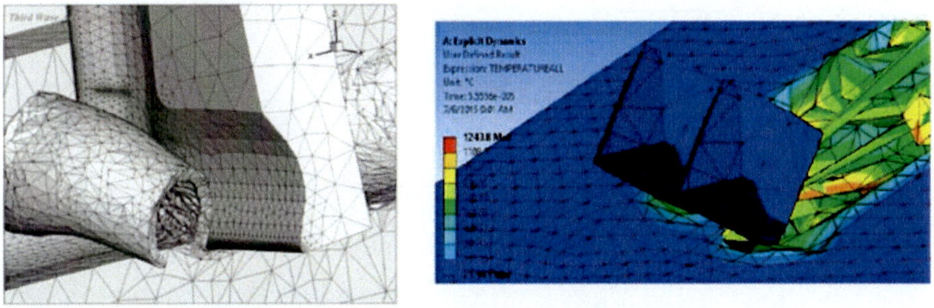

Fig. 5.13 Exemplary 3D models of chip formation in cutting and grinding processes generated in FEM AdvantEdge **a** and Ansys **b** simulation packages [22, 23]

In the FDM method, which is based on the basic heat conduction equation but in differential form, a suitable computational model is built and numerous boundary conditions are defined. The problem is the choice of the heat source distribution and its power [1, 2]. Less applicable is FEM simulation using large spatial elements so-called *voxels* (instead of the classical finite element mesh), which allows to determine the deformation of the workpiece during machining. Modeling without defined mesh (meshless) type SPH (*Smooth Particle Hydrodynamics*) and hybrid type FEM-SPH are developed [1]. The boundary element method (BEM-*Boundary Element Method*) is rarely used [1].

5.4.3 Modelling Using AI Techniques

As mentioned earlier, modeling based on various artificial intelligence (AI) techniques, such as artificial neural networks (ANNs), fuzzy logic (FL), genetic algorithms, *swarm* intelligence, or other learning methods (ML- *Machine Learning*, DL—*Deep Learning* algorithms) are increasingly used in modeling (also as a component of a hybrid model) and optimization of machining processes. They can also contribute to the expansion of empirical modeling capabilities [1].

Figure 5.14 shows the topology of a neural network used to predict the effect of grinding conditions on machining time to surface burn (often related to achievable surface roughness or process efficiency). Expert systems using FL techniques are used for optimization and selection of machining conditions. AI techniques are also successfully used for process monitoring.

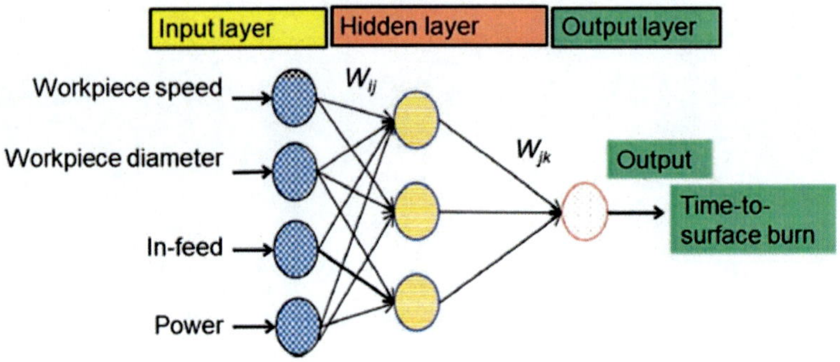

Fig. 5.14 Example of artificial neural network (ANN) architecture with input and output factors for prediction of working time of grinding wheel up to the occurrence of surface burns [24]

References

1. Grzesik W (2017) Advanced machining processes of metallic materials. Elsevier, Amsterdam
2. Melkote S, Grzesik W, Outerio J, Rech J, Schulze V, Attia H et al (2017) Advances in material and friction data for modelling of metal machining. CIRP Ann Manuf Technol 66(2):731–754
3. Arrazola PJ, Ozel T, Umbrello D, Davies M, Jawahir JS (2013) Recent advances in modelling of metal machining processes. CIRP Ann Manuf Technol 62(2):695–718
4. Grzesik W (2020) Modelling of heat generation and transfer in metal cutting: a short review. J Mach Eng 20(1):24–33
5. Klocke F (2011) Manufacturing processes 2. Grinding, lapping, polishing. Springer, Berlin
6. Maradia U, Wegener K (2015) EDM modelling and simulation. In: Electrical discharge machining (EDM): types, technology and applications. Chap. 3, pp 67–121
7. Kozak J (2018) Mathematical modelling of advanced manufacturing processes. Institute of Aviation No. 56, Warsaw
8. El-Hofy H (2005) Advanced machining processes. McGraw Hill, New York
9. Ruszaj A, Grzesik W (2012) Manufacturing of sculptured surfaces using EDM and ECM processes. In: Davim JP (ed) Machining of complex sculptured surfaces. Springer, London Limited pp 229–251
10. Kozak J, Gulbinowicz Z (2011) The computer simulation of shaping in rotating electrical discharge machining. In: Machine learning and system engineering. Springer, London
11. Guitrau EB (2009) The EDM handbook. Hanser Gardner, Cincinnati
12. Kalpakjan S, Schmid SR, Vijay Sekar KS (2014) Manufacturing engineering and technology. Pearson, Singapore
13. Ruszaj A (2017) Some aspects of electrochemical machining process modeling and applications. Adv Manuf Sci Technol 41(4):5–22
14. Niu S, Qu N, Fu S, Fang X, Li H (2017) Investigation of inner—jet electrochemical milling of nickel based Alloy GH4169/Inconel 718. Int J Manuf Technol 93:2123–2132
15. Xu K, Zeng Y, Li P, Zhu D (2015) Study of surface roughness in wire electrochemical micromachining. J Mater Proc Technol 222:103–109
16. Parandoush P, Hossain A (2014) A review of modelling and simulation of laser beam machining. Int J Mach Tools Manuf 85:35–45
17. Robinson M, Jackson MJ (2009) Non-conventional processes: laser micromachining. In: Davim JP (ed) Nano and micromanufacturing. Chap. 6 London, Wiley
18. Jain VK (ed) (2013) Micromanufacturing processes. CRC Press, Boca Raton
19. Aswathy K, Govindan P (2015) Modeling of abrasive water jet machining process. Int J Rec Adv Mech Eng (IJMECH) 4(3):50–71

20. Lebar A, Junkar M (2004) Simulation of abrasive water jet cutting process: part 1. Unit event approach. Model Simul Mater Sci Eng 12:1159–1170
21. Hashish M (1995) Material properties in abrasive water jet machining. Trans ASME J Mater Sci Eng 117:578–583
22. Grzesik W, Niesłony P, Laskowski P (2017) Determination of material constitutive laws for inconel 718 superalloy under different strain rates and working temperatures. J Mater Eng Perf 26(12):5705–5714
23. Kulundzic N, Gostimirovic M, Kovac P, Sekulic M, Savkovic B (2015) Micro cutting simulation of abrasive grains in the grinding process. J Prod Eng 18(2):25–28
24. Pandiyan V, Shevchik S, Wasmer K, Castagne S, Tjahjowidodo T (2020) Modelling and monitoring of abrasive finishing processes using artificial intelligence techniques: a review. J Manuf Proc 57:114–135

Chapter 6
Vibration-Assisted Machining Processes

6.1 Classification of Vibration-Assisted Machining Processes

Vibration assisted machining (VAM) is the kind of assisted machining process which utilizes additional external mechanical energy source of vibrations with a small amplitude of 1–200 μm and frequencies of 0.1–80 kHz (typically about 10 μm and ultrasonic (US) frequency, i.e. within the frequency range of 18–25 kHz). The actuation of vibrations can be realized on the tool or workpiece, and sometimes on the working medium in order to improve material removal process in both conventional machining processes (cutting and abrasive operations) and unconventional machining processes, including EDM, ECM, LBM processes [1–3]. In the case of US vibration, in most systems the vibration of the amplitudes in the range of 1–15 μm and frequencies higher than 25 kHz is generated by piezoelectric vibrator localized in the tool holder, spindle or workpiece holding system [1, 4]. In practice, these assisted processes are denoted by common term *Ultrasonic Assisted Machining* (USAM). It is very important to note that for properly selected machining conditions and vibration parameters the contact between the tool and the workpiece is periodically disrupted and the process mechanics differ substantially from continuous material removal. VAM process is implemented in many cutting operations including turning (VAT), milling (VAM or VAMILL) and drilling (VAD) and in abrasive operations, mainly in grinding (VAG) and polishing (VAP). The main advantages resulting from supplying additional mechanical energy of vibrations are the reduction of componential cutting forces, improvement od surface finish, elimination of burr formation, reduction of tool wear and prolongation of tool life. In turning operations the implementation of vibration assistance is relatively simple because the tool is stationary. Some typical 1D and 2D vibration-assisted processes are presented in Fig. 6.1. In this case (Fig. 6.1c, d) the vibration can be applied as linear vibration in one direction (type 1D) and elliptical vibration in two directions (type 2D).

On the other hand, in the VAMILL-vibration assisted milling process (Fig. 6.1a) the velocity of the cutting edge changes continuously, which causes its trajectory to

W. Grzesik and A. Ruszaj, *Hybrid Manufacturing Processes*, Springer Series in Advanced Manufacturing, https://doi.org/10.1007/978-3-030-77107-2_6

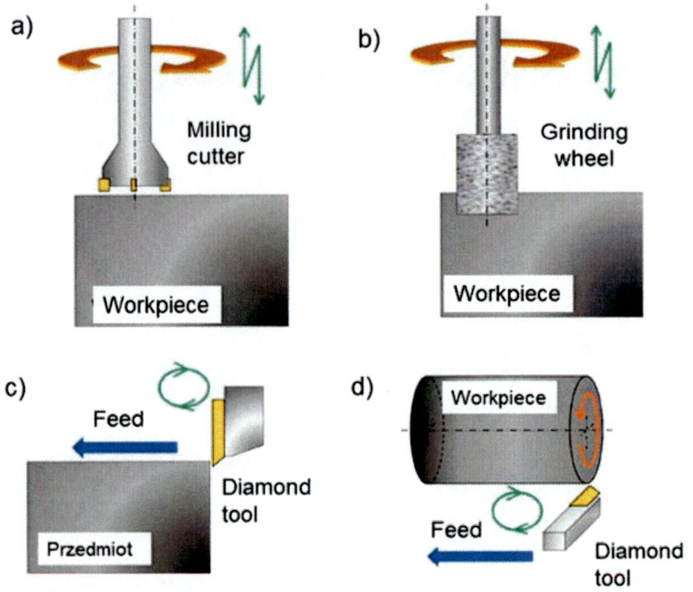

Fig. 6.1 Examples of 1D and 2D vibration-assisted machining processes **a** face milling **b** peripheral grinding , **c** grooving of 2D microstructures, **d** diamond turning [4]

be more complex than in the turning process and the implementation of 1D and 2D vibrations should be based on an accurate analysis of process kinematics (Sect. 6.2.3). In grinding (Fig. 6.1b) additional vibration is added to the grinding wheel, workpiece or working fluid.

It should be noticed that two basic types of vibrations are used, i.e. 1D, which utilizes reciprocal linear vibration (as in Fig. 6.1 a, b) and 2D, which uses circular or elliptical vibrations (as in Fig. 6.1c, d). In addition, 3D type can also be used to produce complex micro/nano-structures [4]. Due to the fact that a stable vibration-assisted process can be performed above the eigenfrequency of the machine tool, i.e. in the range of 200 Hz-1 kHz or the cutting tool, i.e. in the range of 1–10 kHz two diametrically different variants of VAM processes are used. These are [1]:

- Machining process with superimposing low frequency vibrations, typically with frequencies lower than 200 Hz, which is mostly dedicated to reliable chip breaking. This process is termed vibration assisted machining (VAM) or vibration machining and in case of axial vibrations also oscillating machining.
- Machining process with superimposing US vibrations with low amplitudes of 5–10 μm and US frequencies of 15(20)-50 kHz which influences the contact processes on the tool faces and improves the chip formation process. In contrast to the VAM process it is termed ultrasonic assisted machining (UAM).

Vibration-assisted machining systems can also be divided into two basic classes [5, 6]:

- resonant systems, which work at discrete frequencies higher than 20 kHz and produce vibrations with the amplitude lower than 6 μm. The 1D type resonance system is mostly used (80% of all cases) in the VAM processes. The piezoelectric or magnetostrictive transducers/actuators are used in US generators, and the tool is fixed to the concentrator (sonotrode or horn). In such machining arrangement the tool rake face is perpendicular to the vibration direction. The complex 2D vibration movement can be performed by the structural elements of the machine tool (or special vibration device) vibrating with the resonance frequency, e.g. by displacement of the mass centre of diamond tool in relation to the symmetry line for the 1D vibration system [7].
- non-resonant systems, which operate in the frequency range of 1–40 kHz and amplitudes 10-times higher than for the resonance systems. In the case of 2D VAM systems sinusoidal voltage signals are transmitted to piezoelectric transducers/actuators and converted into vibrations with frequencies below the first eigenfrequency of the system. Further, linear displacement of the stack of piezoelectric sensors mounted on the machine tool is, after its excitation, converted into elliptical movement of the cutting edge by mechanical coupling.

6.2 Vibration-Assisted Cutting Processes

6.2.1 *Physical and Technological Effects*

As mentioned above during vibration-assisted machining additional mechanical energy added to the cutting tool or the workpiece is supplied to the cutting zone causing appropriate changes in the energetic balance of the process and its dynamic behavior [1]. Some changes of mechanical loads related to the componential cutting forces and the specific cutting force recorded in turning, drilling and grinding operations are presented in Figs. 6.2a, b and c respectively. It can be noted that additional vibrations cause mechanical loads to decrease.

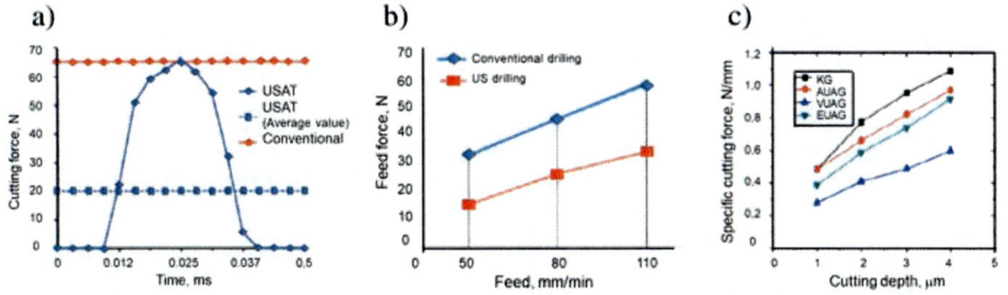

Fig. 6.2 Comparison of process characteristics for conventional and US vibration-assisted machining processes: **a** cutting force in turning; **b** axial force in drilling; **c** specific cutting force in grinding [5]

In general, the influence of vibrations on the process performance can be positive or negative. For instance, chatter, which negatively accelerates tool wear, increases surface roughness and waviness, and dimensional and shape accuracy should be avoided. On the other hand, by adding US vibrations it is possible to intensify some difficult machining operations such as drilling, reaming and tapping of small holes in difficult-to-machine materials including austenitic stainless steels or titanium alloys.

In vibration-assisted machining the vibration direction causes that the cutting parameters (v_c, h_D, a_p) change periodically during the machining process. It can be parallel to one direction of cutting motions (equivalently to one componential cutting force), i.e. vibrations can be generated in tangential, radial and axial directions (see Fig. 6.5a) or it can be bi-directional as a result of composition of two vibrations acting in tangential and axial directions or in tangential and radial directions. These problem are described in detail in Sect. 6.2.3.

Ultrasonic vibrations with the frequency higher than 15 kHz influence the physical mechanism of the cutting process leading to visible improvement of the machinability of the workpiece material. In particular, the excitation of US vibrations with small amplitude of 1(3)-8(10) μm leads to the elimination of build-up edge (BUE) effect, the reduction of chip deformation and strain-hardening effect in the subsurface layer, the modification of friction in the tool-chip contact zone and fatigue material failure at the shear plane. Among the technological effects, the most important include longer tool life, better surface finish (especially in drilling and threading operations), cutting power reduction and a higher dynamic stability of the machine tool-fixture-workpiece-tool system. Positive effects resulting from adding US vibrations were observed in many machining operations including turning, milling, drilling, reaming, tapping, broaching, grinding, superfinishing and polishing operations performed on alloy steels with high ductility and high tendency to BUE formation. In machining of hard materials US vibrations eliminate micro-chipping of the cutting edge and allow obtaining higher dimensional accuracy, surface finish and lower tool wear. As a result, US-assisted machining is utilized when grooving in glass materials and printed circuits. During micro-processes such as EDM, ECM and LBM additional US vibrations lead to the thinner heat affected zone (HAZ), lower residual stresses and practically to the absence of micro-cracks in the subsurface layer [3, 4].

A representative example which confirms the substantial effects of US vibrations is the tapping operation using taps of small diameters of M1-M10 (for metric threads) when the vibration amplitude changed within 3–5 μm and the vibration frequency is in the range of 18–44 kHz. In threading operations performed in heat resistant alloys, titanium and cooper alloys cutting forces and cutting torque were reduced by 25–30%, tool life increased 5-times, productivity increased 3-times and catastrophic tool failure was practically eliminated [1]. However, additional mechanical energy supplied to the contact zone causes that more heat is generated and the contact temperature increases substantially. This effect was observed in drilling operations in Ti6Al4V titanium and 7075-T6 aluminium alloys.

6.2.2 Processes with the Assistance of Low Frequency Vibration (VAM)

Figure 6.3 present a special device for vibration drilling with embedded magnetostrictive transducer for generating torsional vibrations. In general, the sinusoidal movement is used but in ultraprecision machining also elliptical vibrations can be applied [7, 8]. The vibration of the helical drill in the cutting speed direction changes its values and the depth of plastic deformation beneath the machined surface. It is possible to control the thermo-mechanical mechanism of the cutting process in order to reduce the cutting speed in the zone of larger plastic deformation and otherwise to increase it when plastic deformation is less intensive. The final result is that the tool life increases. The condition of the regularity of mechanical loads requires that for sinusoidal vibrations the cutting speed should be $v_c < 2\pi af$.

When vibrations are generated in the feed direction, the depth of cut changes periodically and in the nodal points it decreases down to zero, which means that chip breaking occurs. This effect is easy to obtain in the longitudinal turning when the frequency of axial vibrations is about 5–10% of the workpiece rotation frequency. It is termed turning with oscillating feed. Assuming that chip length between 30(60) and 150(200) mm is the most convenient for evacuation and further transportation from the working space, the required oscillation frequency is only several to a dozen hertz. It should be noted that the corresponding effect causes the chatter to disappear and tool life to increase. The case when the radial vibrations are generated is practically unused because the changes of the depth of cut result in wavy machined surfaces without chip breaking effect.

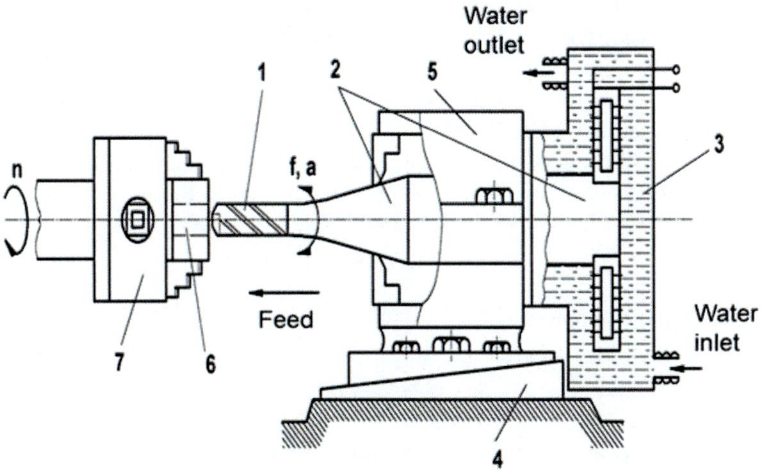

Fig. 6.3 Device for LFV drilling with superimposed torsional vibrations; *1*—drill, *2*—magnetostrictive transducer—concentrator system, *3*—water cooling system, *4*—vertical correction of drill location, *5*—housing, *6*—workpiece, *7*—workpiece fixture [2]

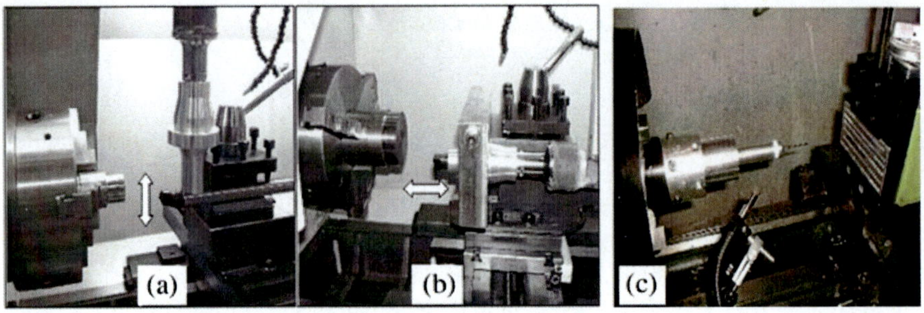

Fig. 6.4 Practical applications of USAM in the turning: **a** in cutting speed direction; **b** in feed direction, and in drilling **c** operations

6.2.3 Processes with the Assistance of Ultrasonic Vibrations (UAM)-Turning, Drilling and Milling

As mentioned above, the US vibration-assisted processes were effectively implemented. in turning (USAT) and drilling (USAD) operations, as shown in Fig. 6.4. In turning vibrations can act in tangential (Fig. 6.4a), axial (Fig. 6.4b) and radial directions (Fig. 6.5a, b) or they can have more complex shape resulting from concurrent summing linear vibrations in the tangential and axial or the tangential and radial directions. In the case when tangential ($v_{vib\text{-}T}$) and radial ($v_{vib\text{-}R}$) vibrations are summed up the resulting elliptical vibrations of the tool tip are obtained (Fig. 6.5c and d) [3, 7]. The elliptical vibrations are widely used in micro and nano-machining processes.

As shown in Fig. 6.6, superimposing vibrations on the tool results in the periodical contact between the cutting edge and the workpiece while the working fluid penetrates easily to the small gap formed during the vibration period (T). It is very important not only in conventional turning and milling operations but especially in micro-drilling and micro-tapping operations because chips are not blocked in the flutes and tool breakage is practically avoided.

During the VAM process the resultant cutting speed v_{ce} changes according to the following formula:

$$v_{ce}(t) = v_c\ A\ \ \omega\ \ \cos(\omega t) \tag{6.1a}$$

where the angular frequency

$$\omega = 2\pi f_{us} \tag{6.1b}$$

Critical speed v_{cr} above which the tool rake face is intermittently in contact with the removed material layer is calculated as

$$v_{kr} = 2\pi f_{us} A \tag{6.1.c}$$

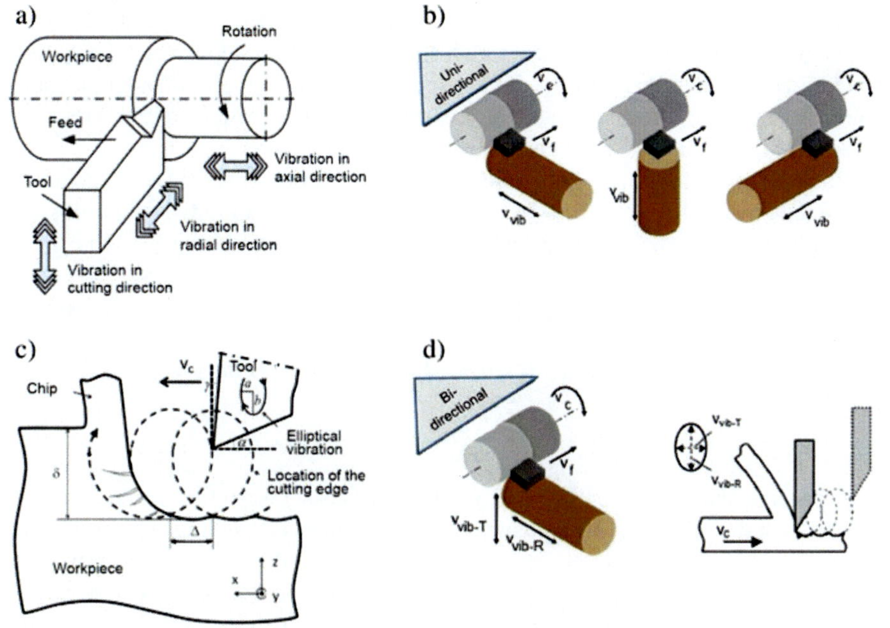

Fig. 6.5 Variants of vibration superimposing in vibration assisted turning operations: (**a** and **b**) linear vibration; (**c** and **d**) elliptical vibration; (**b** and **d**) kinematics of VAT process with 1-axis and 2-axis vibrations [2, 9]

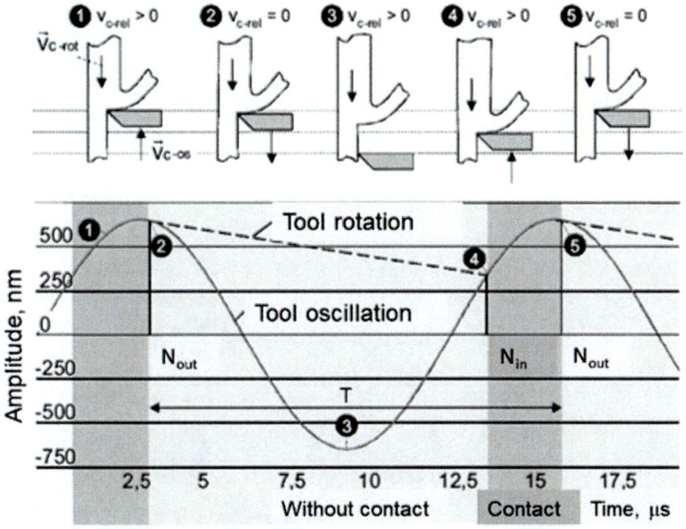

Fig. 6.6 Influence of vibrations superimposed on the tool in the cutting speed direction on the tool-workpiece contact and distribution of vibration amplitude in one vibration period (T) [9]

Hence, for the condition $v_{ce} < v_{cr}$ the tool-workpiece contact is periodically interrupted. In contrast, the process is continuous, i.e. the tool is not disengaged; although the tool tip oscillates periodically, the changes of vibration amplitude are too small. However, this case does not practically exist for common values of the vibration amplitude and frequency when considering the Eq. (6.1b) [9].

Vibration-assisted turning with radial and axial vibrations can be used to generate predefined surface topographies. In particular, in turning with radial vibrations the depth of cut changes according to Eq. (6.2a), i.e.

$$a_{p,res}(t) = a_p + Asin(\omega t) \tag{6.2a}$$

and in the axial direction parallel to the feed movement according to Eq. (6.2b)

$$f_{res}(t) = f + Asin(\omega t) \tag{6.2b}$$

In general, the cutting process of brittle materials with ductile material failure begins when the minimum cut thickness is lower than 1 μm [1]. In such cases more convenient is US-assisted micro-drilling because the axial force is reduced by even 60–70%. Scheme of a UAM system for micro-drilling is shown in Fig. 6.7. The drill vibrating with the amplitude 0.8 μm makes holes of 10 μm diameter in the quartz glass with the penetration rate of 0.05 μm/s and the ductile regime is achievable for the layer thickness of about 0.05 μm.

The principle of USAM process dedicated to predefined surface micro-texturizing is presented in Fig. 6.8. The idea is similar to that based on controlled changes of

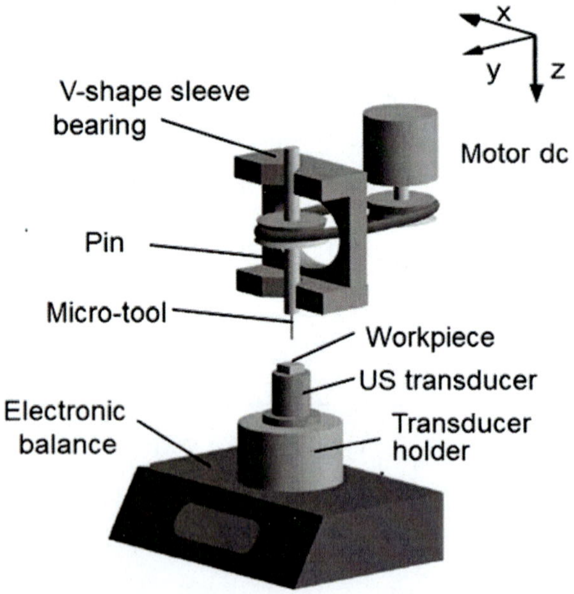

Fig. 6.7 Scheme of UAM device for micro-drilling operations [1]

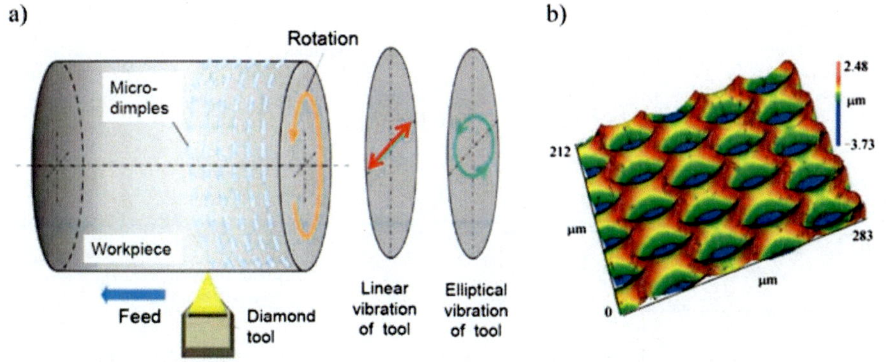

Fig. 6.8 Applications of 1D and 2D turning operations with US vibration assistance of surface micro-texturizing **a**; typical surface structure produced with superimposing elliptical vibrations **b** [4]

the surface topography resulting from additional vibrations in the radial and axial directions. It is possible to form micro-dimples during the discontinuous cutting process when using 1D type vibrations. On the other hand, the addition of 2D type vibrations generated by a special spindle allows producing complex micro-structures resulting from changes of the cutting speed and the depth of cut. That means the generated elliptical vibrations can be effectively utilized for producing predetermined surface textures on the part surfaces, as exemplarily shown in Fig. 6.8b.

The working principle of modified milling head which generates US vibrations with 18 kHz frequency on the cutting edges of the milling cutter is presented in Fig. 6.9. For this purpose a stack of piezoelectric sensors in the form of piezo-quartz

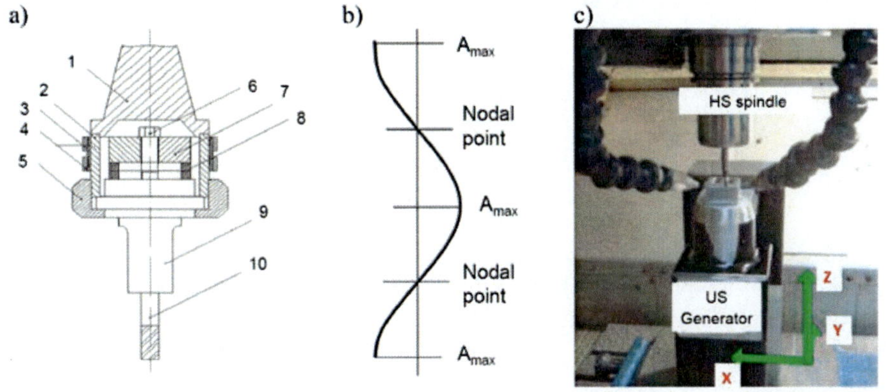

Fig. 6.9 Structure of the head for VAMILL operation **a** working principle **b** and view of real VAMILL process **c**. Description: 1—standard Weldon-type holder, 2—bush, 3—textolite bush, 4—collector rings, 5—nut, 6—crew, 7—pressure bush, 8—piezo-quartz rings, 9—horn, 10—milling cutter [6, 8]

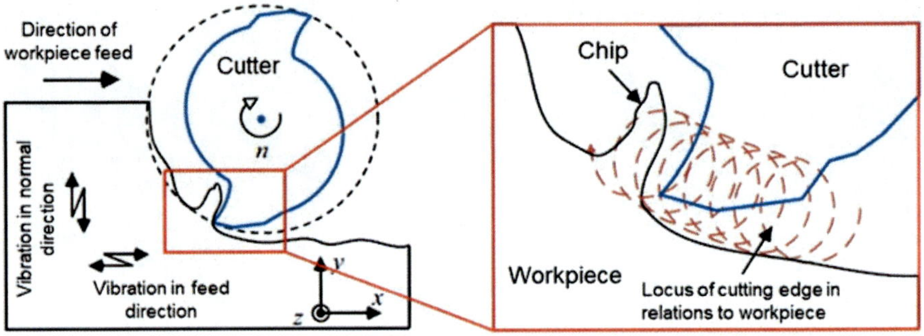

Fig. 6.10 1D and 2D-type kinematics of slab milling operations with superimposed vibrations [8]

rings was embedded into the head (position 8 in Fig. 6.9). Vibrating milling cutter was fixed in the head using standard Weldon-type holder (Fig. 6.9c). It is important to note that the vibrating system works under the resonance conditions with the nodal point located on the cutter teeth, as shown in Fig. 6.9b.

In general, vibration-assisted milling operations can be divided into two main groups [1, 10]:

- 1D type operations (1D VAMILL) with the assistance of vibrations in the feed direction (FVMILL) or in the perpendicular direction (cross-feed vibration milling-CFVMILL), as shown in Fig. 6.10.
- 2D type operations (2DVMILL) with simultaneous vibrations applying in these two directions.

Figure 6.10 presents a schematic diagram of vibration-assisted slab milling with the previously defined vibration directions for both 1D and 2D variants. The *x,y,z* coordinate system includes the workpiece feed direction in *x*-axis, the cross-feed direction in *y*-axis and the axial depth of cut (a_p) in *z*-axis.

Assuming that vibrations are added to the workpiece, the coordinates of the vibration trajectory can be determined as follows:

$$\begin{cases} x_v = Asin(2\pi f_x t + \emptyset_x) \\ y_v = Bsin\left(2\pi f_y t + \emptyset_y\right) \end{cases} \tag{6.3}$$

where: *A* and *B* are the vibration amplitudes, f_x i f_y are the vibration frequencies, and ϕ_x and ϕ_y are phase angles in the *x* and *y* directions respectively.

Next computation step includes the calculations of the relative displacements x_i and y_i of the cutter tips to the workpiece in VAMILL operations. The most important for a successful vibration milling is the definition of appropriate pattern of the tool-workpiece separation (TWS) resulting from the rapid periodic interruption of constant tool-workpiece contact. There are three different mechanisms which lead to the separation phenomenon [6, 7]. They are denoted by types I, II and III and illustrated in Fig. 6.11.

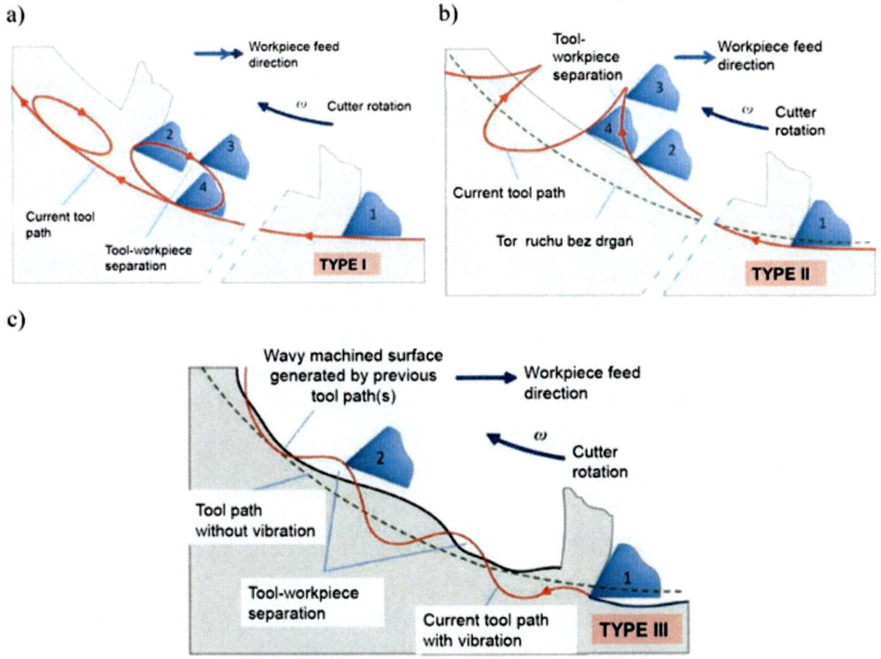

Fig. 6.11 Three separation types of cutter tip in slab milling operations: **a** type I (TWS I), **b** type II (TWS II), **c** type III (TWS III) [5, 8]

As shown in Fig. 6.11a the separation of I type (TWS I) is present in the current trajectory of the cutter tip when the tangential component of the vector of relative velocity $\overrightarrow{v_e}$ between the tool and the workpiece is directed oppositely to the tool rotation. As a result, the tool tip is located behind the workpiece surface. In position *1* the tangential component of $\overrightarrow{v_e}$ vector is greater than zero, which means that the tool-workpiece contact exists, but subsequent movement of the tool tip to point *2* causes that it decreases to zero and separation occurs. In the next position *3* when this component is negative, i.e. in the opposite direction to the cutter rotation, the cutter tip disconnects from the workpiece surface. It should be noticed that during the separation phase, the direction of the $\overrightarrow{v_e}$ vector changes from zero to negative and returns to positive. Consequently, in point 4 the tool-workpiece contact occurs again.

As shown in Fig. 6.11b the separation of II type (TWS II) occurs in the current trajectory of the cutter tip tool when its radial displacement (measurement direction of the cut layer thickness) resulting from added vibration is greater than the instantaneous cut layer thickness (CLT). This movement period is equivalent to the tip displacement from point *1* to point *2*, i.e. the radial displacement is equal to the instantaneous value of CLT. The next phase of tip movement to position *3* causes that the vibration displacement exceeds the instantaneous value of CLT and the cutter tip completely separates from the workpiece surface. Starting from point *4* the tip continuously engages again into the workpiece.

As shown in Fig. 6.11c the separation of III type (TWS III) is related to the generation of wavy machined surface (not to a relatively smooth surface as in turning) which also influences material removal process. This type of dynamic machining process is termed *regeneration of surface waviness* [2]. It can be seen in Fig. 6.11c that the tool tip in the next pass encounters a previously generated wavy surface and removes a material layer with periodically variable thickness. The newly formed surface is again wavy and it is possible to break the tool-workpiece contact in some overlapping regions. Correspondingly, discontinuous chips are formed in these regions.

Although the three types of the tool-workpiece separation can be kinematically selected in VAMILL operations, in practice, they can also be generated simultaneously. The kinematic details of the tool-contact conditions for milling operations with additional vibrations are discussed in [1].

6.3 Vibration-Assisted Grinding and Polishing Processes

During grinding excitation of vibrations is typically performed in the grinding wheel-workpiece system or during the emulsion flushing of the wheel in order to remove the particles of strongly deformed material from pores on its working surface. Correspondingly, vibrating particles of the cooling fluid penetrate easily to the contact zone. However, vibration-assisted grinding is a relatively new hybrid machining process so in practice the superposition of conventional grinding with US vibrations is applied [9].

Figure 6.12a presents the vibrating system with a piezoelectric exciter connected by a rod with the vibrating table. The system with vibrating workpiece is used in peripheral face grinding operations. On the other hand, Fig. 6.12b shows a vibrating

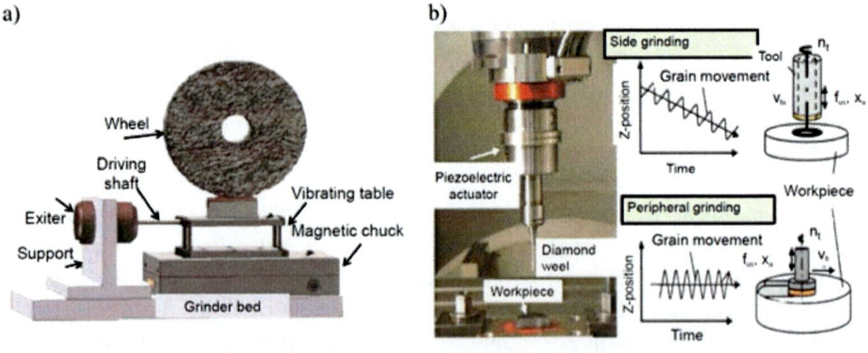

Fig. 6.12 Device for producing US vibration on the workpiece in face peripheral grinding operation **a** [5] and vertical peripheral and face grinding using diamond wheel **b** [3, 9]

system where the relative tool-workpiece vibrations are obtained through excitation in the tool holder system (it also is possible in the spindle or on the workpiece itself). As shown in Fig. 6.12b this solution is practically utilized for side and peripheral grinding using diamond wheel. The vibrations drastically change the conventional scratching type of grains action into a sinusoidal penetration with impact character. Many advantages resulting from additional vibrations are reported in grinding operations of hard carbides and ceramics, including reduced mechanical loads, better surface quality or subsurface integrity [9].

In the 2D type vibrating device shown in Fig. 6.13 the US elliptical vibrations were obtained by superposition of bending and longitudinal vibrations with a small amplitude of several micrometres. The piezoelectric plate as the patch-type vibrator (PZT) was clamped to the face surface of the metal block which is elastically connected with the fixture. The workpiece in the form of a thin plate is fixed on the top of the vibrating block. The obtainable vibration amplitudes are only of 0.4 μm due to relatively large mass of the vibrating system. In addition, the influence of workpiece mass on the vibration amplitude and its frequency is unavoidable. This vibrating system was tested similarly as in standard scratch tests using a diamond tip and obtained results suggest that, under US vibrations, the critical depth of cut increases from 0.3 μm to about 1.5 μm [12].

Figure 6.14a presents the principle of rotary ultrasonic machining (RUM), also termed rotary ultrasonic assisted grinding (RUAG), which is a combination of classical ultrasonic machining (USM) and grinding using cylindrical metal bonded diamond tool [13]. In practice, the grinding process can be implemented on a 3–5-axis CNC centre with the capability of milling operations but also special ULTRASONIC type machine tools with 5-axis CNC systems are produced [6]. The wide spectrum of possible applications of RUM technology include the machining of ceramic materials, quartz glass, sintered carbides, hard steels, composite materials and precious stones (ruby and sapphire). In addition, HSK actuator can be integrated with 5-axis

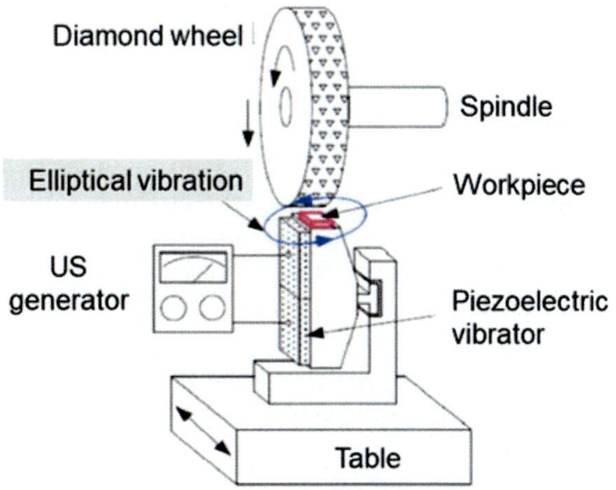

Fig. 6.13 Device for producing elliptical vibrations on the workpiece in ultra-precision grinding process [6, 12]

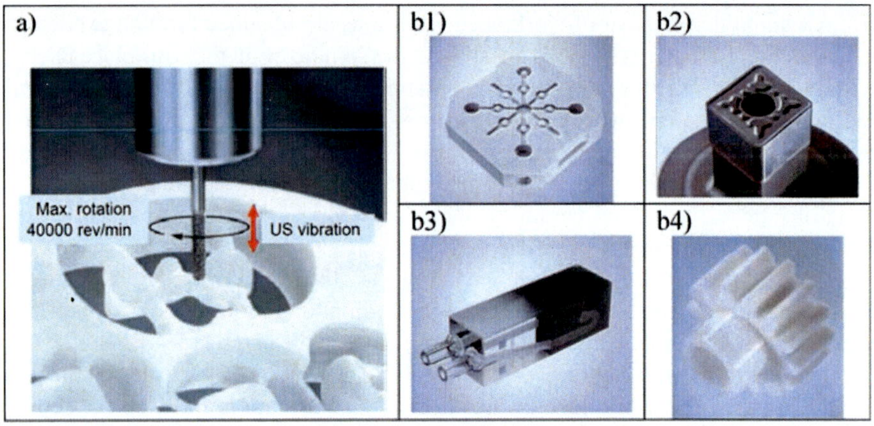

Fig. 6.14 RUAG process under HSC milling conditions using rotating tool with bonded diamond abrasive grains **a** and application examples (**b**1–**b**4) [1, 6]

milling to perform HSC milling operations. The frequency of axial vibrations ranges from 16.5 to 30.5 kHz and the maximum rotational speed is about 40 000 rev/min. The productivity of URM process is even 5-times higher than traditional USM process and achievable surface roughness is Ra < 0.2 μm.

This innovative grinding technology is already implemented in various areas of optical, biomedical, automotive, aircraft and aerospace techniques. For instance, as shown in Fig. 6.14b, it is successfully applied in fabrication of assembly plates made from oxide ceramics (b1), precision punch with a copied form of the carbide tool insert (b2), a glass cell with miniature holes for blood analysis (b3) and miniature gear made from zirconia ceramics (b4).

US vibration assistance is also applied in other micro-abrasive processes such as polishing and/or lapping processes. US assisted lapping is used in producing microstructures with high aspects ratios of even larger than five in hard and brittle materials, it is also reported that ultrasonic assisted polishing machine allows the reduction of the Rz roughness parameter of polished surfaces on micro-aspheric dies and molds made from tungsten carbides in the range of 8 nm [3, 9]. US vibration-assisted polishing system with 2-axis vibrator designed for micro-aspheric molds is shown in Fig. 6.15a. A small polishing tool in the form a pencil of 2 mm in diameter and made from stainless steel (Fig. 6.15b) is mounted on a 3-axis or 5-axis controlled work table and 2D vibrations are generated by piezoelectric actuators. During lapping a polyurethane abrasive paste is smeared on the spherical tip of the shaft. Another solution shown in Fig. 6.15c is that the lapping tool is integrated with a magnetostrictive transducer and the exciter consists of four coils with 200 turns made from a wire of 0.16 mm in diameter. The elliptical US vibrations are generated by appropriately setting the two pairs of opposing coils. The vibrator operates at the frequency of 9.2 kHz [14].

Figure 6.16a presents a 3D vibration polishing system for generating 3D

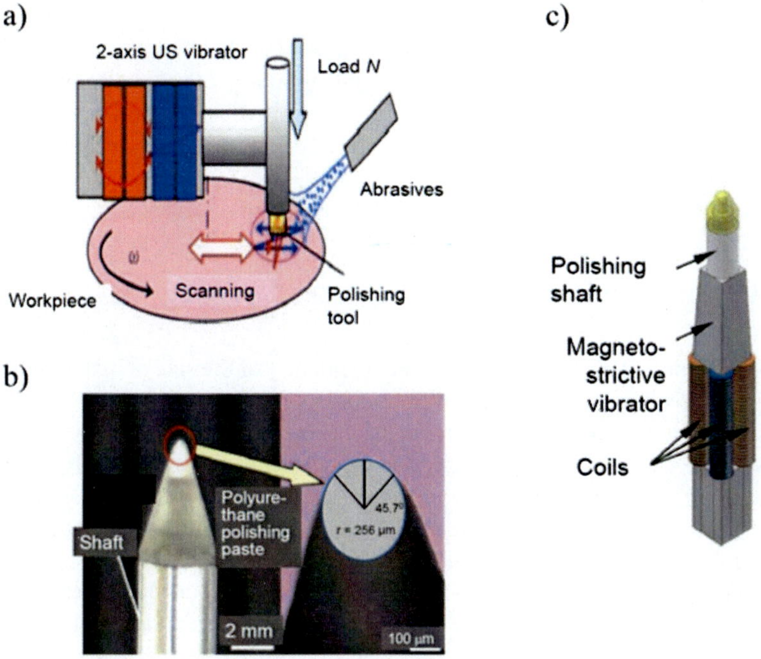

Fig. 6.15 Scheme of 2D US vibration-assisted polishing **a** and the view of pencil-type polishing pin **b** with integrated magneto-strictive exciter **c** [3, 11, 14]

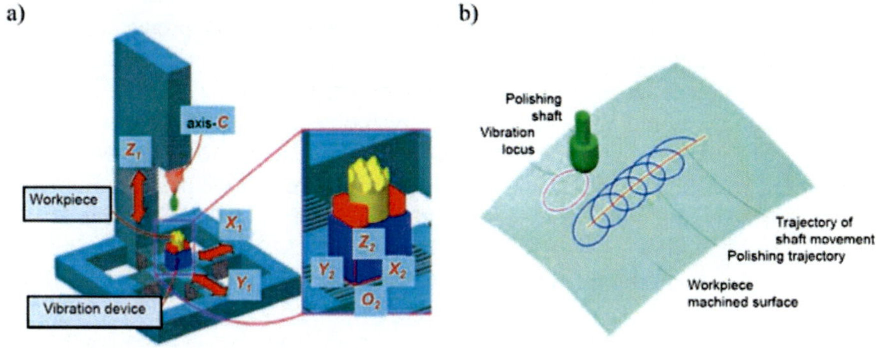

Fig. 6.16 Scheme of 3D US vibration-assisted polishing with 4-axis CNC control system **a** and the view of trajectory of polishing pin with elliptical vibrations **b** [7]

microstructures based on a 4-axis machine tool and a 3D vibration device. As shown in Fig. 6.16a, the four movement axes, i.e. X_1, Y_1, Z_1 and C, and three axes of the vibrating device, i.e. X_2, Y_2 and Z_2, are controlled. The concept is based on the utilization of elliptical vibrations similarly as for rotational surfaces presented in Fig. 6.8. This system is ideal for polishing of small surfaces with greater curvatures because an accurate feed and high frequency vibrations are produced. In particular, the tool is mounted in the spindle and the workpiece is fixed to the 3D vibration device. As a result (Fig. 6.16b), a precision continuous polishing trajectory as a combination of the machining path and the vibration locus is generated. Apart from a tight load of the polishing shaft to the structured surface, the polishing trajectory should be evenly distributed on the workpiece surface.

6.4 Vibration-Assisted EDM Processes

Hybrid EDM processes can be divided, independently of the size of machining zone, into two groups, i.e. assisted and pure (combining) processes. In the case of hybrid EDM processes the main material removal mechanism, i.e. electric discharges, is assisted by external energy source, as for instance by vibrations applying on the working electrode, the workpiece or the working fluid (dielectric). On the other hand, material removal results from the integration of various mechanisms, as for instance in simultaneous EDM/ECM (SEDCM) process or in EDM-assisted grinding (EDAG). In these processes the effect of process synergy occurs but all componential processes take part separately in the material removal and generation of the surface integrity.

During the process, the working electrode vibrating at US frequency moves towards the workpiece. Dielectric fluid is supplied to the interelectrode gap and material removal mechanism is based on the electric discharges. The EDMUSM process is predominantly dedicated to micro-hole drilling and die sinking due to improved evacuation of particles of the melted material and erosion products from the discharge zone and the interelectrode gap. Under some favourable process conditions the cavitation phenomenon can occur in the interelectrode gap and it can be activated by additional US vibrations. These specific phenomena include three phases, i.e. formation, growth and implosion of the gas bubbles. The implosion of gas bubbles causes rapid local significant changes of the temperature and the pressure. In these micro-areas the temperature raises to about 5000 K (its gradient is in the range of 10^7 K/s) and corresponding pressure increases to about 100 MPa. Under these extremal working conditions in the interelectrode gap material is removed more effectively [15].

A scheme of the US vibration-assisted EDM (EDUSM) process along with US vibration control system, when the vibrations are applied to the electrode-tool, is shown in Fig. 6.17. In this case the EDM machine is equipped with additional vibrating head and the generator of US vibrations. However, as mentioned above, US vibrations can also be applied to the workpiece (Fig. 6.18) and the working fluid (Fig. 6.19). It is possible to obtain, depending on the vibration transducer used, the vibration frequency in the range of 18–25 kHz or a higher value of about 40 kHz. However, the vibration frequency does not influence the process performance markedly. Typically, the optimal working conditions during the EDUSM process are achieved for the electric impulse time lower than 30 μs, current during the impulse

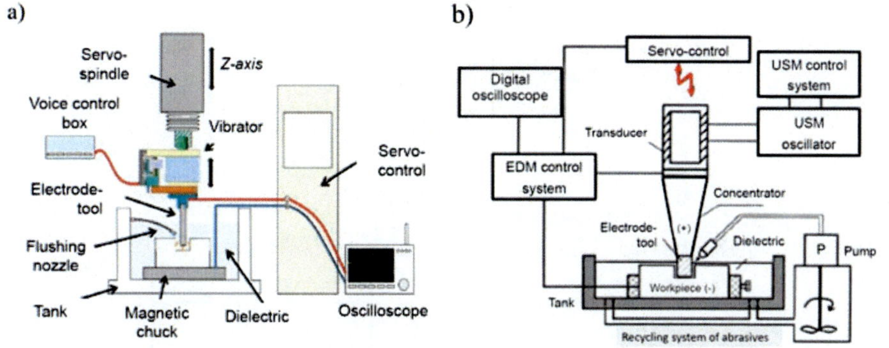

Fig. 6.17 Scheme of vibration-assisted EDM process with vibrations superimposed on the tool-electrode **a**, US vibration control system **b** [16, 17]

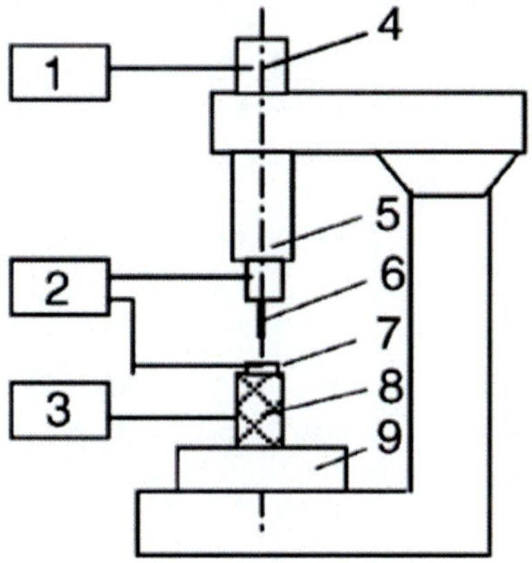

Fig. 6.18 Scheme of vibration-assisted EDM process with vibrations superimposed on the workpiece [3]. Symbols: 1—motor controller, 2—impulse generator, 3—US generator, 4—dc motor, 5—spindle, 6—tool-electrode, 7—workpiece, 8—piezoelectric transducer, 9—working table

lower than 5A, vibration amplitude equal to 0.1–0.2 of the working gap width and a small machining area [9, 18]. It was documented that US vibrations influence the process performance when using hydrocarbon dielectric under finish machining conditions. It can be stated that in macro-scale the positive effect of US vibrations in the EDM process can be observed for smaller machined surfaces and finishing passes. It is very important in fabricating 3D micro-parts with the dimensions of about 100 μm, and dimensional tolerance lower than 1 μm and surface roughness Ra < 0.5 μm made from special materials. In the case when the US transducer is adhered to the workpiece (Fig. 6.18) the efficiency of micro-EDM process can increase 8 times [9, 19].

It can be noted in Fig. 6.19 that in order to vibrate the working fluid, an US transducer is attached to the dielectric tank and vibrations are applied by a horn tip.

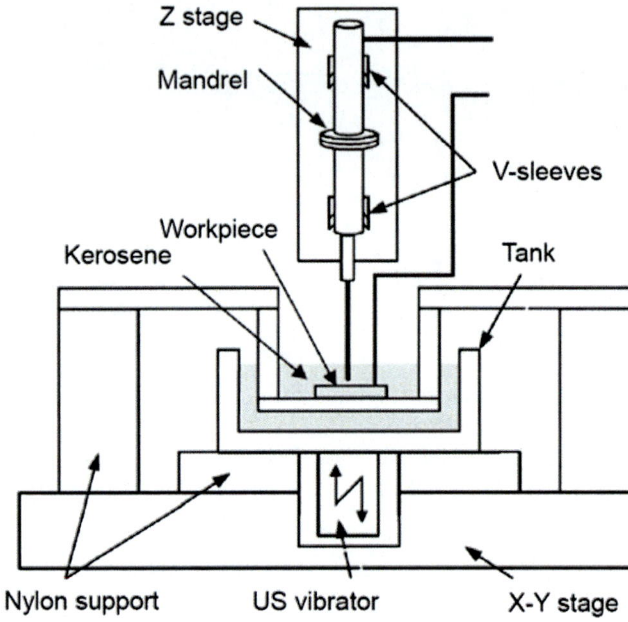

Fig. 6.19 Scheme of vibration-assisted EDM process with vibrations superimposed on dielectric fluid [3]

A small circular hole is made in the horn and when electrode is fed through this hole, US vibrations can be applied to the working fluid without the contact between the tool electrode and the vibrating horn. It is a good method to circulate debris, improve process performance and achieve higher material removal rate.

Direct US vibration of the working electrode or the workpiece is considered as the optimal strategy for improving dielectric flushing in the interelectrode gap and stabilize the process in high aspect ratio structures. The second effect of the periodic changes of the interelectrode gap can be achieved using the EDUSM system with vibrating working fluid shown in Fig. 6.19. As a result, the process speed can be significantly enhanced and moreover, as illustrated in Fig. 6.20, the tool feed is undisturbed and complex structures can be reproducibly fabricated [9].

6.5 Vibration-Assisted ECM Processes

Hybrid ECM processes can be divided, taking into account their participation in the material removal, into two groups, i.e. assisted and pure (combining) processes. In the case of hybrid ECM processes the main material removal mechanism, i.e. electrochemical reactions resulting in electrochemical dissolution, is assisted by external energy source, as for instance by vibrations applied on the working electrode, the

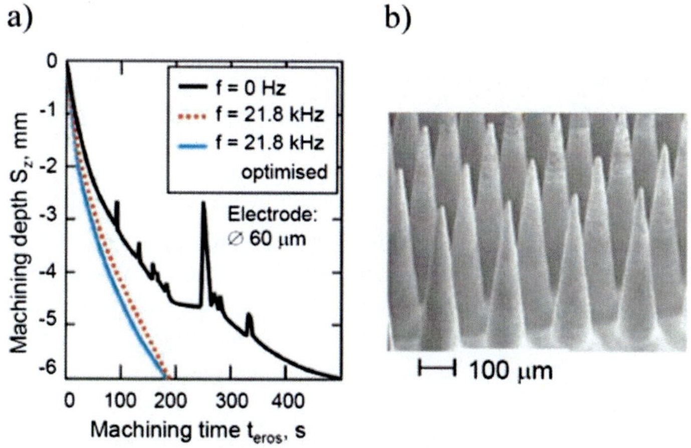

Fig. 6.20 Conditions of US assisted micro-EDM process **a** and 3D structure produced when using vibrating electrode of 60 μm diameter [9]

workpiece or the working fluid (electrolyte), and laser heating of the removed material layer. On the other hand, for pure ECM processes material removal results from the integration of various mechanisms, as for instance in simultaneous EDM/ECM (SEDCM) process or in ECM-assisted grinding (ECAG). In these processes the effect of process synergy occurs but all componential processes take part separately in material removal and the generation of the surface integrity.

In ECM processes, the hydrodynamic conditions in the interelectrode gap are very important because, during the process, a desired portion of electrolyte of an appropriate temperature and saturation with the products of electrochemical reactions should be supplied continuously to the machining zone to maintain efficient flushing. These conditions are necessary for stabile machining process during which electrolyte absorbs the heat generated by the passage of current and short-circuits do not occur. It is practically not possible in ECM processes performed on larger surfaces with areas of dozens cm^2.

These effects are achieved in both electrical and mechanical aspects as pulsed ECM variant, termed pulsed electrochemical machining (PECM) [1, 3, 20]. The problem of applying both additional impulse voltage and vibrating working electrode cannot be decisive in the generation of a new hybrid process when additional energy source is considered as a hybridization criterion. It is only optimization of the energy distribution versus time which causes that higher machining accuracy and uniform surface integrity but lower material removal rate is obtained. It is an important reason that the entire ECM process is performed in a rough pass as a classical die -sinking process with maximum productivity and finish passes under PECM conditions in order to obtain the desired accuracy and/or surface integrity. It should then be noted that the integration of low frequency tool-electrode oscillations and pulsed voltage allow achieving an optimal distribution of the process energy in terms of process efficiency, dimensional accuracy and surface quality [1, 3, 18, 20, 21].

Exemplarily, the pulsed ECM process can be based on the superimposition of US vibrations with the frequency of about 20 kHz and the wave length of $\lambda = 0.072$ m. i.e. substantially longer than the inter-electrode width (typically $s \ll 1$ mm), on the working electrode or the workpiece [21]. From this point of view, the detailed analysis of the process phenomena in the working gap, the secondary effects, such as cavitation and energy absorption caused by US vibrations should be considered [21]. In ECM the effect of cavitation concerns the generation of pulsed bubbles which appear as a result of local rapid pressure increases lower than the cohesion forces. The localization of cavitation zones can be roughly determined based on the analysis of the static pressure (p) based on the following condition:

$$p \le p_n \tag{6.4}$$

where: p_n – pressure of the saturated electrolyte vapour under defined flow conditions.

In general, the cavitation effect occurs when the density of US wave achieves the limiting values, i.e. cavitation threshold [22, 23]. Its value depends on the fluid type, wave frequency and the presence of microscopic impurities and gas particles. It means that the mixture of electrolyte and dissolution products intensifies the cavitation process. It was documented that the bubble implosion effect causes some rapid local changes of the temperature and pressure in the working zone. The principle of US-vibration-assisted ECM process (USECM) and the general scheme of machine tool are presented in Fig. 6.21 a and b respectively.

It should be noticed that physical phenomena in the machining zone (Fig. 6.22b) are analogous as in ECM die-sinking presented in Fig. 6.21a. The application of US vibrating ball-end working electrode improves hydrodynamic flow conditions and causes di-passivation of the machined surface. As a result, electrochemical

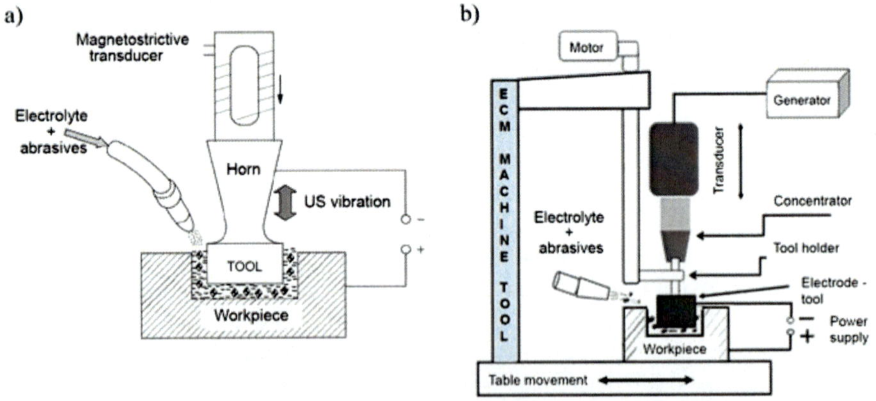

Fig. 6.21 Principle of USECM method **a** and scheme of machine tool for ECM process **b** [22, 23, 25] (www.home.iitk.com.in)

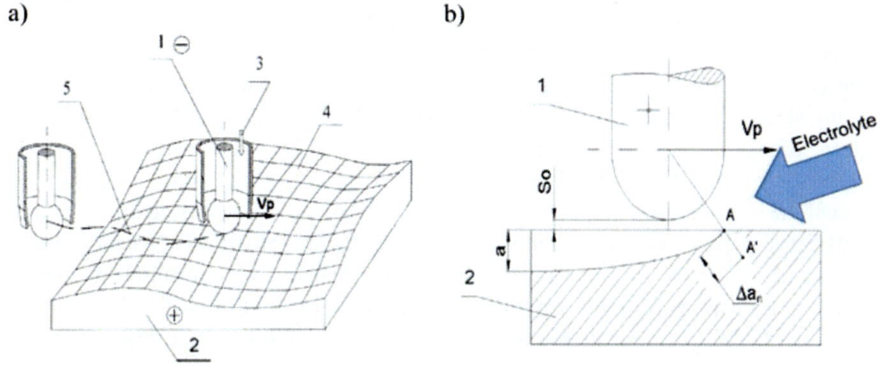

Fig. 6.22 Scheme of USECM planing operation **a** and the illustration of machining zone **b**; 1—universal working electrode, 2—workpiece, 3—electrolyte supply, 4—machined surface, 5—trajectory of electrode movement; symbols: S_o—initial interelectrode gap, a—thickness of removed material, v_p—velocity of ball-end working electrode movement [18, 24]

dissolution becomes more intensive because the width of interelectrode gap as well as the thickness of removed material increase in comparison to traditional ECM process [11–13]. Correspondingly, surface roughness decreases distinctly. For instance, minimum width of the interelectrode gap and low roughness parameter Ra of about 0.8 μm are recorded for the machining power of about 70 W. In general, the desired shape of the machined surface is generated in USECM sinking process but both milling and planing variants can also be utilized [6, 13]. The final shape of the machined surface is produced by copying the working electrode path as shown in Fig. 6.22a.

The principle of the vibration-assisted ECM die-sinking process (PECM-*pulsed electrochemical machining*), shown in Fig. 6.23, is that a low frequency oscilla-

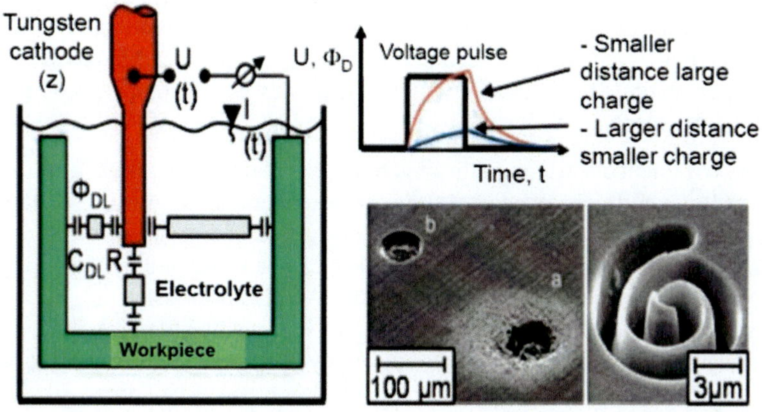

Fig. 6.23 Principle and exemplarily application of PECM process [9]

tion superimposed on the tool-electrode along with ultra-short voltage pulsation are applied to the interelectrode gap. As a result, during time intervals when he interelectrode voltage is equal to zero and the gap width is higher than the working gap, the flowing electrolyte flushes the products of electrochemical reactions and absorbs the generated heat. When the current is switched on, the ECM process occurs again but in a fresh electrolyte, and the ECM process becomes more stable and reliable. The main effect of the combination of voltage pulsation and the oscillating tool-electrode is that the working gap is reduced to about 10–50 μm [9]. Therefore, this hybrid method is especially effective in fabricating complex microstructures with the dimensions higher than 5 μm, as for instance miniature punches (image in Fig. 6.23) or cutting inserts. The application of ultra-short voltage pulses shifts the machining accuracy to the nanoscale, for instance the achievable accuracy of 100 nm was obtained for 500 ps voltage pulse duration [9]. Comparatively, holes (a) and (b) and in Fig. 6.23 were made with the pulse durations of 5 ms and 100 ns, and the accuracy of hole (b) is visibly higher.

6.6 Industrial Applications of Vibration-Assisted Machining Processes

As shown previously in Fig. 2.7, the most machining processes and also some metal forming processes can be improved by supplying the additional vibration energy. However, currently only some processes such as vibration assisted turning, grinding and polishing can be applied in massive production. The application of other conventional machining processes including milling and drilling is limited to prototype setups. On the other hand, vibration-assisted unconventional processes, as for example EDM, are generally in the testing phase. In general, there are many technical limitations in the wider application of vibration-assisted material removal processes, although they can substantially enhance the performance of main processes.

The idea of the utilization of vibrations, mainly ultrasonic vibrations, is developed since the sixties of the twentieth century. During above six decades a lot of technological and economic benefits of VAM/USAM processes are documented including the following aspects:

- the improvement of surface quality and reduction of machining time. These effects are predominantly observed in cutting operations including turning, drilling and milling, but also in grinding and polishing operations of difficult-to-machine materials. Figure 6.24 illustrates how the surface can be smoothed by the application of vibration with a controlled amplitude in VAMILL process.
- the enhancement of machining processes of hard and brittle materials, such as ceramics and glass. These materials are widely used in precision, optical, medical and dental industries. A good example is the innovative RUG process.

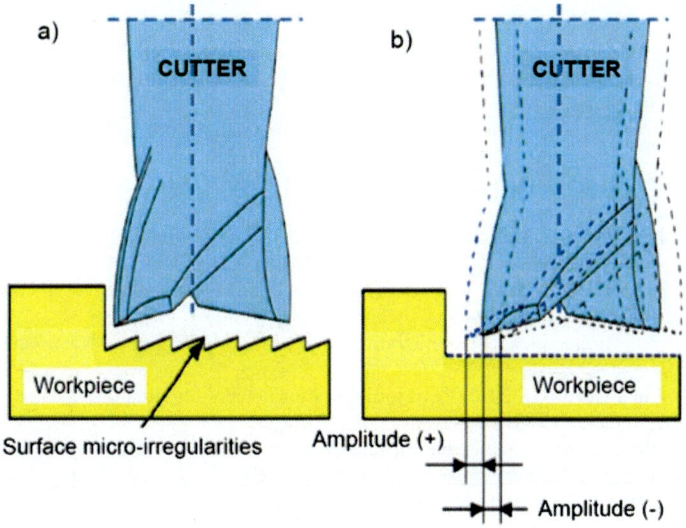

Fig. 6.24 Effect of VAM process on surface irregularities in face milling operation [8]

- the improved efficiency of cooling/lubricating (working) fluids due to oscillated tool movement. This aspect corresponds not only to ecological but also to economic benefits resulting from longer tool usage.
- the fabrication of complex 3D micro-textures which is highly important for higher part functionality due to improved tribological condition between contacted surfaces [14]. USM processes are more and more often used in the production of miniature die and moulds, ultra-precision optical parts and complex 3D microstructures.
- VAM processes become an alternative for conventional machining processes of hard steels, hard non-ferrous materials and alloys and special alloys, as for instance nickel based Inconel 718 alloy. A rapid development of resonant and non-resonant vibrating devices and machines is observed.
- Vibration-assisted EDM process is widely used in automotive, aircraft and aerospace industry in order to facilitate the technological indexes. Similarly, vibration-assisted ECM is also the leading process in these industry branches but dedicated to conductive workpiece materials.

References

1. Luo X, Qin Y (2018) Hybrid machining. Academic Press, London
2. Grzesik W (2017) Advanced machining processes of metallic materials. Elsevier, Amsterdam
3. Chavoshi S, Luo X (2015) Hybrid micro-machining processes: a review. Prec Eng 41:1–23

4. Xu S, Kuriyagawa T, Shimada K, Mizutani M (2017) Recent advances in ultrasonic-assisted machining for the fabrication of micro/nano-textured surfaces. Front Mech Eng 12(1):33–45
5. Kumar M, Kanmani S, Vamsi P, Venugopal A (2014) Vibration assisted conventional and advanced machining: a review. Proc Eng 97:1577–1586
6. Brehl D, Dow T (2008) Review of vibration-assisted machining. Prec Eng 32(3):153–172
7. Wang G, Liv B, Zheng Q, Zhou H, Liu Z (2019) Polishing trajectory planning of three-dimensional vibration assisted finishing the structured surface. AIP Adv 9:015012-1–0115012-14
8. Chen W, Huo D, Shi Y, Hale M (2018) State-of-the-art review on vibration-assisted milling: principle, system design, and application. J Adv Manuf Technol 97:2033–2046
9. Lauwers B, Klocke F, Klink A, Tekkaya E, Neugebauer R, McIntosh D (2014) Hybrid processes in manufacturing. CIRP Ann Manuf Technol 63(2):561–583
10. Ramkumar J, Vardhman A (2020) Vibration assisted milling. Chapter 4 Micromanufacturing Lab IIT Kanpur
11. Zheng L, Chen W, Huo D (2020) Review of vibration devices for vibration-assisted machining. J Adv Manuf Technol 108:1631–1651
12. Liang Z, Wang X, Wu Y, Xie L, Jiao L, Zhao W (2013) Experimental study on brittle-ductile transition in elliptical ultrasonic assisted grinding (EUAG) of monocrystal sapphire using single diamond abrasive grain. Int J Mach Tools Manuf 71:41–51
13. Suzuki H, Hamada S, Okino T, Kondo M, Yamagata Y, Higuchi T (2010) Ultraprecision finishing of micro-aspheric surface by ultrasonic two-axis vibration assisted polishing. CIRP Ann Manuf Technol 59(1):347–350
14. Brinksmeier E, Glaebe R, Schoenemann L (2012) Review on diamond-machining processes for the generation of functional surface structure. CIRP J Manuf Sci Technol 5(1):1–7
15. Ichida Y, Sato R, Morimoto Y, Kobayashi K (2005) Material removal mechanism in non-contact ultrasonic abrasive machining. Wear 258:107–114
16. Lin YCh, Yan BH, Chang YS (2000) Machining characteristics of titanium alloy (Ti-6Al-4V) using a combination process of EDM with USM. J Mater Proc Technol 104:171–177
17. Tsai M, Fang C, Yen M (2018) Vibration-assisted electrical discharge machining of groove in a titanium alloy (Ti-6Al-4V) J Adv Manuf Technol 97:297–304
18. Kozak J (2018) Mathematical modelling of advanced manufacturing processes. Science library of the institute of aviation no 56 Warsaw
19. Gao C, Liu Z (2003) A study of ultrasonically aided micro-electrical-discharge machining by the application of workpiece vibration. J Mater Proc Technol 139:226–228
20. Cardoso P, Davim JP (2012) A brief review on micro-machining of materials. Rev Adv Mater Sci 30:98–102
21. Banks CE, Compton RG (2013) Voltametric exploration and applications of ultrasonic cavitation. Chem Phys Chem 4(2):169–178
22. Ruszaj A, Zybura M, Zurek R, Skrabalak G (2003) Some aspects of the electrochemical machining process supported by electrode ultrasonic vibrations optimization. Proc IMechEng, Part B: J Eng Manuf 217:1365–1371
23. Ruszaj A, Zybura-Skrabalak M, Skoczypiec S, Zurek R (2001) Electrochemical machining supported by electrode ultrasonic vibrations. In: Proceedings ISEM XIII, Spain Bilbao, pp 953–964
24. Skoczypiec S, Ruszaj A, Czekaj J, Zybura-Skrabalak M (2003) Primary investigations on USECM-CNC process. In: McGeough JA (ed), Proceedings 18th international conference on computer-aided products engineering CAPE, Edinburgh, Wiley, London pp 367–375
25. El-Hofy H (2005) Advanced manufacturing processes. McGraw Hill, London

Chapter 7
Media-Assisted Machining Processes

7.1 Classification of Media-Assisted Machining Processes (MAM)

It should be reminded that in the general classification of hybrid shaping processes (Chap. 1.3) [1, 2], in Group *I, assisted processes (subgroup I.A)* and *mixed/combined processes* (subgroup *I.B*) are distinguished. In the case of conventional and unconventional machining, the most important is the creation of hybrid processes according to the principle *I.A*, including the assistance of various cutting methods with fluid and gaseous media (MAM–*Media Assisted Machining* or FAM–*Fluid Assisted Machining*) [3]. The types of cooling and lubricating media used in cutting and abrasive machining processes are shown in Fig. 7.1.

The use of liquid (fluids and gases) and solid media requires, due to their specific properties, different supplying methods, such as flooding or minimum quantity of lubricant (MQL) or coolant and lubricant (MQCL) in the machining medium. The characteristics and effects of different cooling and lubrication strategies (dry cutting, MQL and cryogenic cutting) in the machining process are shown in Table 7.1. They are inextricably linked to the trend to increase the *sustainability* of the machining process [1, 4]. In the early 1970s, when tool coatings began to be used on an industrial scale, it seemed possible to expand the use of dry cutting on a large scale. However, the required progress in surface engineering of wear-resistant thin-film materials was not achieved for four decades and in the early 1990s, *near-dry machining,* commonly referred to as machining with MQL—(MQL–*minimum quantity lubrication),* began to be introduced into industry as a new sustainable machining strategy. It allowed a significant reduction in the use of coolants/lubricants, but further barriers were encountered in improving the machining process, e.g. tool life and surface quality/technological surface layer condition. Like traditional cooling lubricants, manufactured aerosols also have a negative, although lesser, impact on the environment and the personnel's health. This strategy does not work for difficult-to-machine

W. Grzesik and A. Ruszaj, *Hybrid Manufacturing Processes*, Springer Series in Advanced Manufacturing, https://doi.org/10.1007/978-3-030-77107-2_7

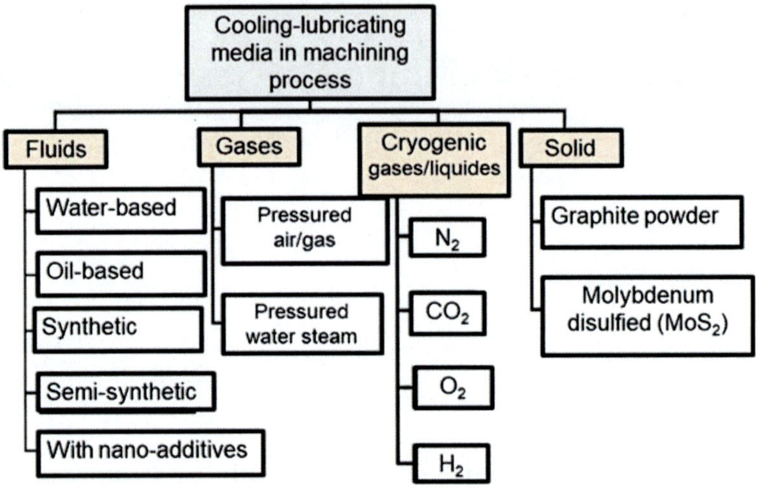

Fig. 7.1 Classification of cooling and lubrication media applied in machining processes [4]

Table 7.1 Characteristics and applications of different cooling and lubrication strategies in machining processes [1]

Type of impact		COOLING AND LUBRICATION STRATEGIES				
		Wet/flooding (emulsion/oil)	Dry (pressurised air)	MQL (oil)	Cryogenic (LN2)	Hybrid (LN2+MQL)
Primary	Cooling	good	poor	marginal	very good	very good
	Lubrication	very good	poor	very good	marginal	very good
	Chip disposal	good	good	marginal	good	good
Secondary	Machine tool-cooling	good	poor	poor	marginal	marginal
	Workpiece- cooling	good	poor	poor	good	good
	Dust/particles-control	good	poor	marginal	marginal	good
	Part quality	good	bad	marginal	very good	very good

materials such as hardened steels, superalloys, or composite materials. It became necessary to introduce another sustainable machining strategy based on the concept of dry machining with rapid cooling effect using gaseous and liquid cryogenic media (liquid nitrogen LN_2, cooled CO_2, cooled air) individually or in a hybrid mode, e.g. LN2 combined with MQL (Table 7.1).

7.2 Physical, Tribological and Technological Effects

Various cutting fluids (cooling fluids–CFs) are used in machining practice, such as water, synthetic fluids, water–oil emulsions, synthetic and natural oils, cryogenic fluids and gases as well as methanol. Their use is dictated by the need to control the cutting process and select it to the required quality and productivity. Traditional cooling lubricants are designed to remove heat from the cutting zone and reduce heat generation by reducing friction in the machining zone. The reduction of friction and the intensity of plastic deformation of the material results in a decrease in mechanical load and power demand, as well as providing longer tool life [1–3].

For this reason, hybrid MAM/FAM machining processes are designed to provide the required process efficiency in cases of extremely difficult machining conditions. This is particularly important when machining materials with poor machinability. Mechanical, thermal and tribological changes are observed in the assisted processes, and the combined effect of these three interactions is shown in Fig. 7.2. Positive effects in terms of improved machinability and increased productivity are obtained by applying highly pressurized CFs (*HPC–High Pressure Cooling*), which mainly concerns machining of difficult-to-machine materials—Ni and Ti alloys, ceramics, composite materials and corrosion-resistant steels [1, 6]. Chip breaking control is also very important (Sect. 7.3.1).

On the other hand, the thermal effect on the tool, consisting in rapid cooling of the machining zone (Sect. 7.4.2) causes changes in the properties of the tool and workpiece material with simultaneous intensive heat dissipation from the machining

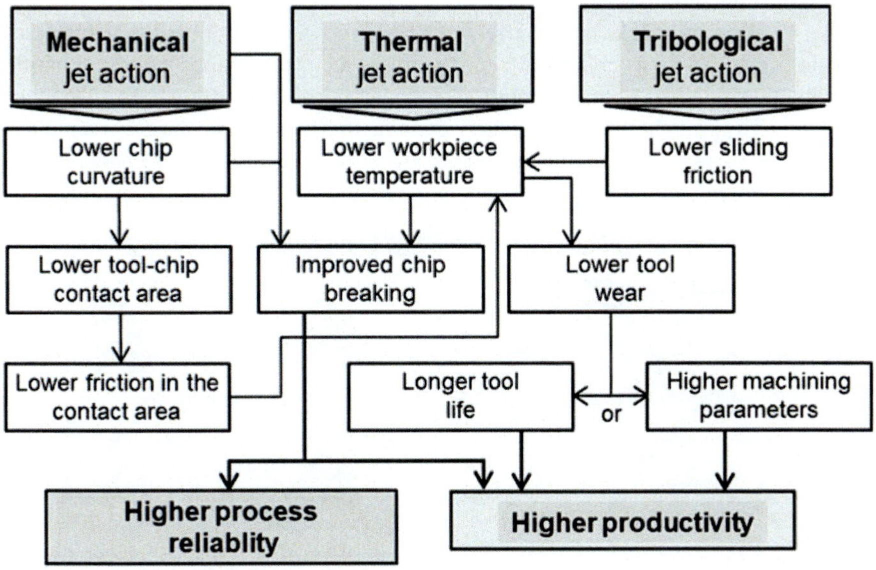

Fig. 7.2 Influence of high-pressured fluid jet on the performance of cutting process [5]

zone, beneficial for the tool life, chip morphology and functionality of the machined surface (e.g. increase in fatigue strength, resistance to corrosion and abrasive wear). The documented effects of LN_2 on the cutting process include reduced plastic deformation, reduced friction, reduced abrasive wear on the flank face and crater formation on the rake face. These have been demonstrated in the machining of titanium alloys, superalloys (e.g. Inconel 718 alloy), magnesium alloys, composites based on metal matrix, hard steels, cast iron, and medical memory-shape materials [1, 7]. For example, in cryogenic machining of titanium alloys, tool life can increase tenfold and machining productivity can double.

7.3 Fluid Media-Assisted Machining Processes

7.3.1 Fluid Media-Assisted Cutting Processes

Among the media-assisted cutting processes, the supply of high-pressure cooling lubricants (HPC–high pressure cutting) has demonstrated significant practical importance, mainly for improving the machinability of difficult-to-machine materials—Ni and Ti alloys, ceramics, composite materials and corrosion-resistant steels. Research in this area has been conducted for several decades, and the beginning is attributed to Iscar company, which introduced a high-pressure coolant supply system for cutters and grooving tools called *Jetcut* in the early 1990s. The spread of the HPC machining concept is credited to Sandvik Coromant, which has applied it to various variants of turning, milling and drilling tools.

As it is commonly known, the effect of directing a jet of fluid under appropriate pressure in the gap between the chip and the rake face is to produce a hydrodynamic wedge that changes the contact conditions in the chip-tool zone (Fig. 7.3a). The fluid flow rate and its pressure are the most important factors influencing the chip formation process, i.e. its shape and breaking effect, the tool wear mechanism and its final

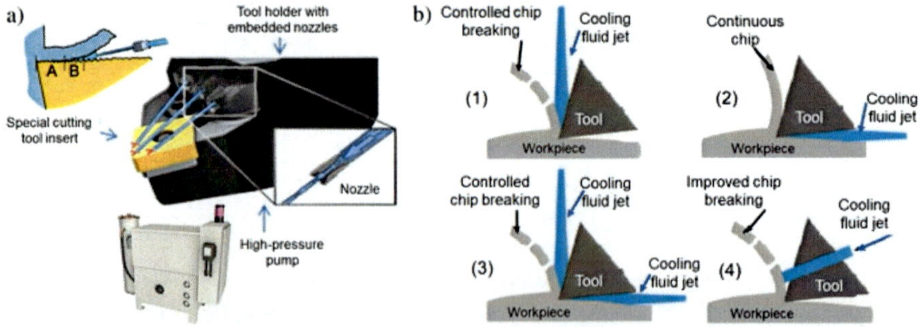

Fig. 7.3 Principle of the generation of hydrodynamic effect in the tool-chip contact zone **a** [5] and possible cases of the influence of fluid jet on chip formation mechanism **b** [9]

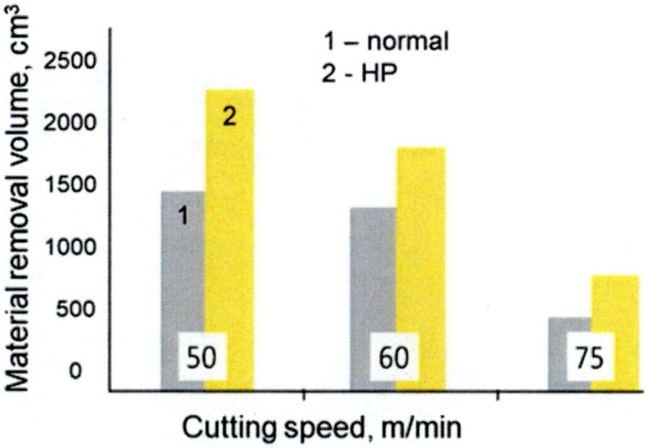

Fig. 7.4 Comparison of machining productivities in the machining of Ti6Al4V titanium alloy [5] (www.sandvikcoromant.com)

durability. Changes are also observed in the structure of the machined material and the chip. These are associated with significant mechanical, thermal and tribological changes (the mentioned reduction of the chip-tool contact length in Fig. 7.3a) of the process. It can be concluded that the hydrodynamic effect is similar the use of grooved cutting inserts (using RCT tools) [1]. For this reason, this method of supporting the process with process media is also referred to as (*high-pressure lubricoolant supply*) [4]. In tool systems for turning (Fig. 7.3a) and boring it is recommended to supply CF at a pressure of 70–80 bar for stainless steels and 150–200 bar for duplex stainless steels and HRSA alloys [8]. The recommended flow rate is 20 l/min.

In the case of very high pressure (above 300 bar), the terms high pressure jet assisted machining (HPJAM) or high pressure water jet assisted machining (HPWJAM–*High Pressure* WJAM) are used [5]. The principle of the HPC process and the possibilities of controlling the chip breaking are shown in Fig. 7.3b. It should be noted that the chip breaking improves considerably when the liquid jet is delivered through a channel in the cutting insert directly onto the upward curling chip (case *4* in Fig. 7.3). Fig. 7.4 shows the increase of MRR for various cutting speeds resulting from the application of HPC system.

Figure 7.5 shows the change in chip shape in grooving Inconel 718 alloy as the cooling lubricant jet pressure is increased to 80 and 300 bar respectively. In Fig. 7.5b, it can be observed that increasing the HPC pressure to 80 bar causes the elimination of tangled chips that seriously disturb the cutting process in automated machining. A similar effect was obtained in turning titanium alloy [10].

Since the practical application of the HPJAM method is limited to rough machining, an extension to finishing is proposed by pulsating CF supply at high pressure in a short time interval (Δt_1 in Fig. 7.5b) when reliable chip breaking is required [11]. In the subsequent, but longer, time interval Δt_2 the pressure is significantly reduced to the value necessary to maintain cooling and lubrication effects in the machining zone. This promotes the economical use of cutting energy. This action is justified by the significantly higher specific cutting energy when the energy

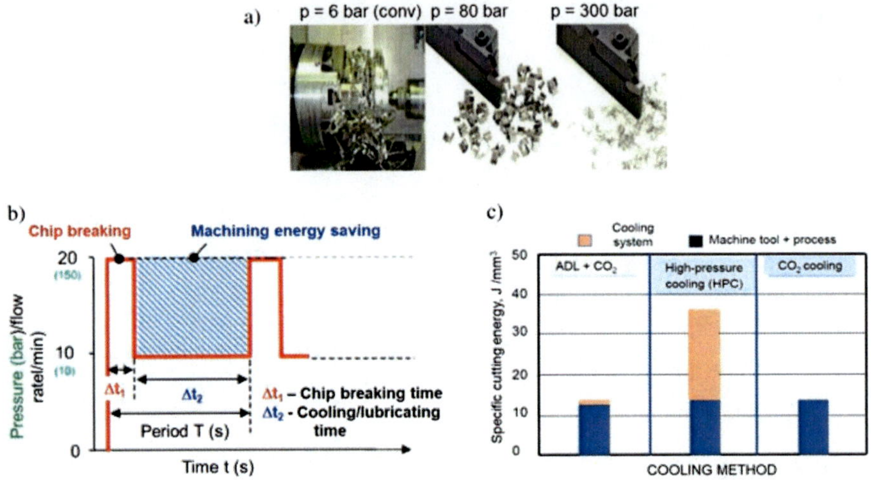

Fig. 7.5 Improvement of chip breakability in grooving of Inconel 718 alloy by supplying cutting fluid (CF) under pressure of 80 and 100 bar **a**; machining conditions: cutting tool material–CBN, CF–emulsion 8%, $v_c = 250$ m/min, $f = 0.1$ mm/rev, $b = 6$ mm, $a_p = 5$ mm (conventional cooling) and 10 mm (HPC) [2], **b** diagram of the working cycle of machine tool for pulsated process (HPJAM), **c** comparison of the specific cutting energy (SCE) in the machining of Ti6Al4V titanium alloy ($v_c = 70$ m/min, $f = 0.2$ mm/rev, $a_p = 1.5$ mm [11]

consumption for powering the high-pressure pump supplying CF to the cutting zone is taken into account (Fig. 7.5c).

Based on many years of experience using HPC technology, the primary suitability is for cutting difficult-to-machine materials in the aerospace industry, e.g. turbine components. The low thermal conductivity of these materials results in accelerated and often catastrophic tool wear when a proper cooling strategy is not used. It has been established that, in general, the cutting temperature can be reduced by 40–50% [4, 6]. This is confirmed by the record of temperature changes under different machining conditions (the temperature in dry machining increases almost 4 times when CF supply is stopped) in Fig. 7.6. For this reason, it is possible to increase the cutting speed up to the HSM limit in turning and milling of titanium alloy Ti6Al4V and Inconel 718. According to literature data, the material removal rate in machining Ti6Al4V titanium alloy increases by 50% at the same cutting speed (Fig. 7.4) and additionally by 20% as an effect of increasing the cutting speed [5]. Under the HPJA machining conditions, as in Fig. 7.6, a 30% reduction in tool temperature compared to conventional cooling is achievable with a 350% reduction in CF flow rate [6].

As a result of the relaxed thermal conditions, tool wear in the machining of Inconel 718 superalloy, also of a large group of steels, can be reduced by up to 350% in finishing operations. The wear curves shown in Fig. 7.7 confirm the high sensitivity of wear to cooling method and conditions in the machining of duplex stainless steel and Inconel 718 alloy.

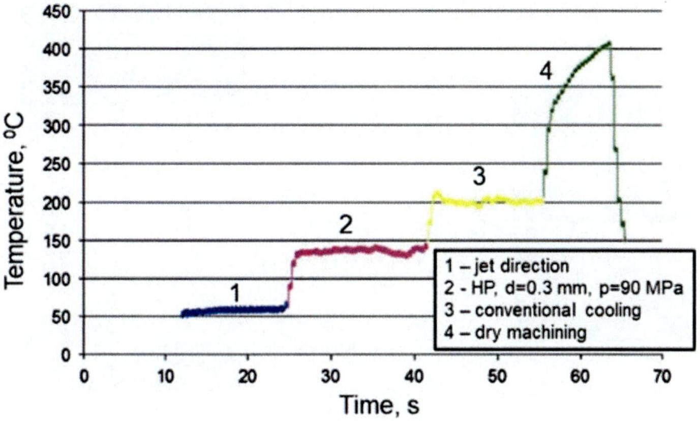

Fig. 7.6 Comparison of temperature evolution during one cooling cycle in the machining of Inconel 718 alloy. Machining conditions: $v_c = 50$ m/min, $f = 0.224$ mm/rev, ap = 2 mm [6]

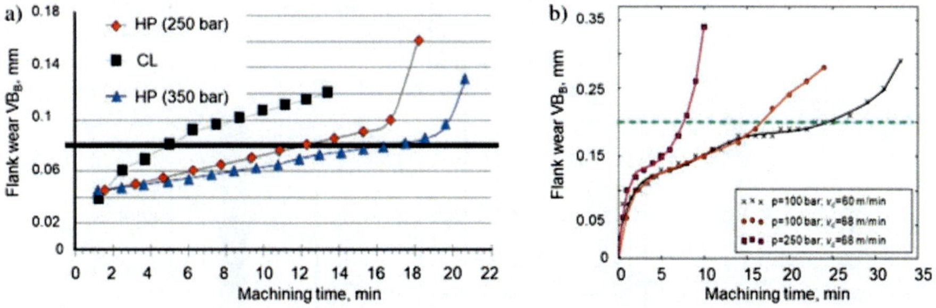

Fig. 7.7 Influence of CF pressure on the tool wear in the machining of duplex steel X2CrNiMo22-5 for $f = 0.15$ mm/rev, $a_p = 0.5$ mm **a** and Ti 555–3 titanium alloy for $f = 0.1$ mm/rev, $a_p = 1.5$ mm **b** [10, 12]

7.3.2 Fluid Media-Assisted EDM Process

The dielectric plays a fundamental role in the EDM sinking process because its properties determine the width of the inter-electrode gap and thus the course and effect of the discharge—i.e. the volume of material removed, the size of the craters created by a single discharge. The relationship between the surface area and depth of a crater is important in terms of material removal rate and surface layer properties. Intuitively, it can be assumed that a greater crater depth contributes the formation of a rougher surface. On the other hand, a larger crater area means a lower energy density, i.e. a smaller thickness of the altered layer, more favourable conditions for the removal of molten or vaporised material, and in this way into an increase in material removal rate. One way to improve the dielectric properties is to use a dielectric-powder mixture (Fig. 7.8). Scheme of the device for the PMEDM—*powder-mixed*

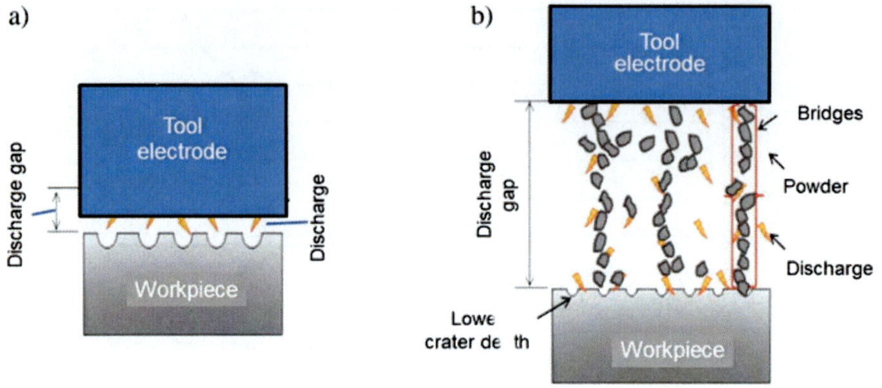

Fig. 7.8 Principles of conventional EDM process **a** and PMEDM **b**

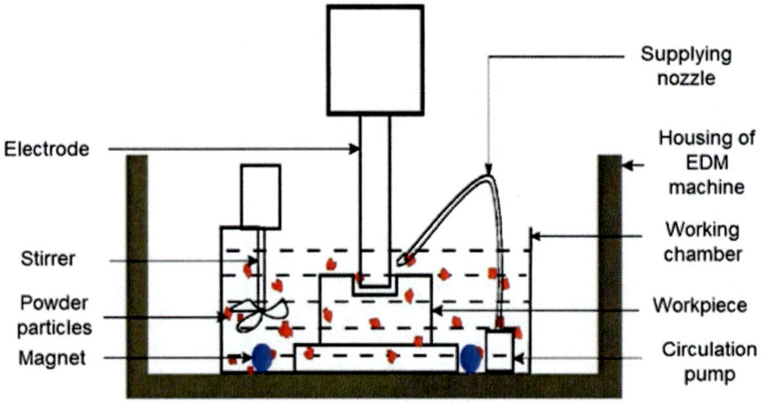

Fig. 7.9 Scheme of producing of dielectric-powder mixture in the PMEDM process [13]

electrical discharge machining is shown in Fig. 7.9.

The addition of powder to the dielectric (Fig. 7.10) causes a change in the distribution of the electric field in the gap, which, as mentioned earlier, has a significant effect on the course and effects of the discharge. Of course, the basic condition for technological success is the even powder concentration distribution. If this condition is not met, the addition of powder usually worsens the technological indices of the process. In Fig. 7.10a the concentration of discharges in one localized zone can be observed.

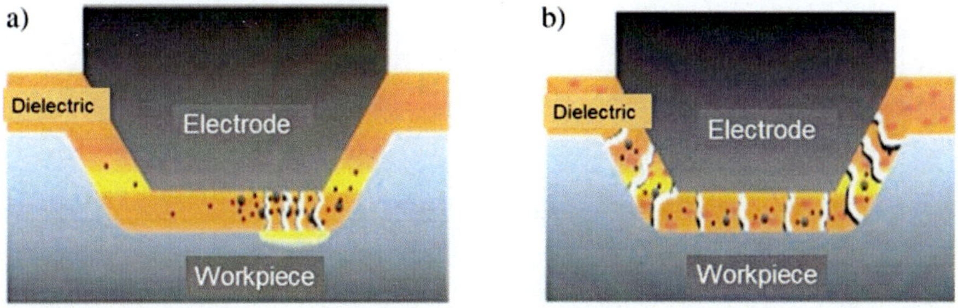

Fig. 7.10 Visualization of the working zone of EDM processes for inhomogenous **a** and homogeneous **b** dielectric-powder mixture

Table 7.2 Influence of powder grade on the width of interelectrode gap [15]

Powder material	Width of interelectrode gap (μm)
Without powder	10–15
Graphite	45–50
Silicon	32–27
Aluminum	120–160
Crushed glass	10–15
Silicon carbide (SiC)	80–90
Molybdenum disulfide (MoS_2)	44–48

Process conditions: cathode material – Cu, workpiece material – SkH54 steel, current amplitude (I_p) – 1A, dielectric flow rate – 15mls^{-1}; time between impulses (t_o) – 11 μs; single impulse time (t_i) – 7.5 μs

The influence of the powder material added to the dielectric on the width of the inter-electrode gap is given in Table 7.2 based on literature data [14–17]. It should be noted that for most powders the inter-electrode gap width increases significantly up to a concentration value of about 10 g/litre. A further increase in concentration has only a slight effect on the gap width.

Figure 7.11 shows the effect of the addition of SiC and Al_2O_3 particles in the dielectric on the machining performance (MRR) taking the impulse time (a) and current amplitudes (b) as parameters.

The main effects of using a mixture of dielectric and powder in the EDM process are:

- possibility of efficient and stable machining in dielectrics based on deionised water,
- possibility of obtaining a good quality surface layer by reducing the surface roughness Ra to about 0.6 μm and reducing the number of micro-cracks.

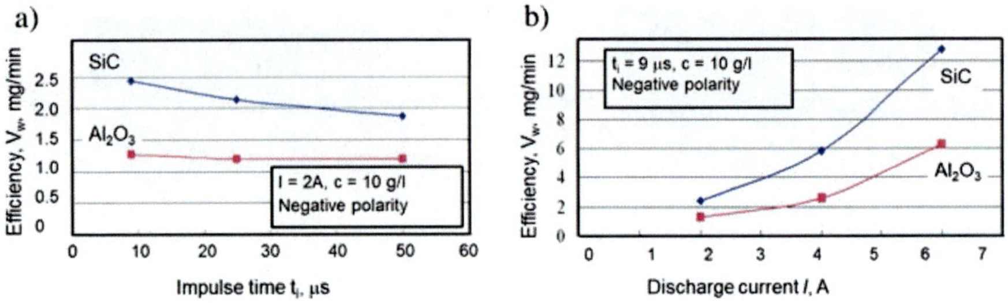

Fig. 7.11 Influence of impulse time **a** and current amplitude **b** on the process productivity when using SiC and Al_2O_3 powder with grain size of 17 μm [15]

Studies [18, 19] have shown that it may be technologically interesting to use gaseous dielectrics [18] based on the fact that pure gas is as good an insulator as contaminated oil. However, gas as a dielectric must be supplied under high pressure to ensure that it flows through the machining zone at high speed. Only at this supplying regime can the gas jet remove the vaporised and molten metal from the workpiece surface and restore the insulating properties of the inter-electrode gap.

Due to the high pressure and high velocity of the gas flow, this machining variant can conventionally be classified as a hybrid method as—*EDM assisted by a high-pressure gas jet*. In addition, due to the need to eliminate short circuits, this process is best supported by ultrasonic vibrations. This means that it is a hybrid EDM process assisted simultaneously by a high-pressure gas jet and ultrasonic vibrations of the workpiece (Fig. 7.14).

The applicability of EDM machining in a gaseous atmosphere was verified for the case of machining of shaped cavities (3D) using a copper tube electrode moved along a suitably designed trajectory, analogous to EDM milling [18]. In this case, gas is supplied through a hole in the working electrode (Fig. 7.12).

Fig. 7.12 Scheme of EDM process in gaseous atmosphere [18]

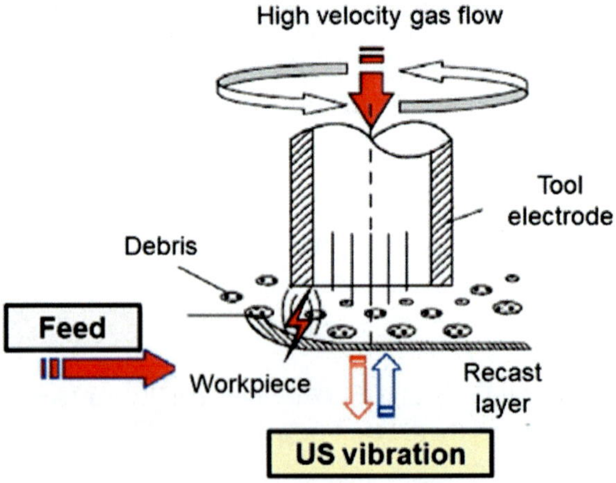

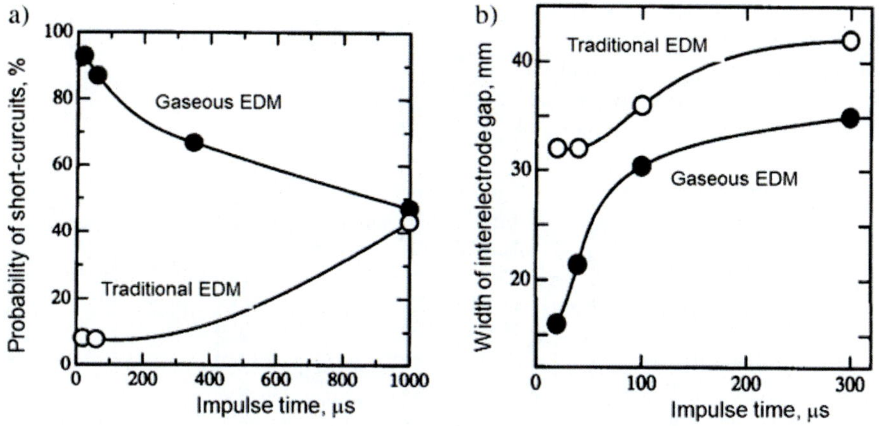

Fig. 7.13 Influence of impulse time on the probability of short-circuit occurrence. **a** and the width of working gap, **b** process conditions: ignition voltage of $U = 280$ V, thickness of electrode wall of 0.5 mm, current amplitude of 20 A, gas used pressured air at 0.5 MPa [19]

A comparison of some technological indicators obtained for hydrocarbon dielectric liquids and air is shown in Fig. 7.13 [19]. Much lower electrode wear should be noted compared to conventional machining in oil, but, the material removal rate is lower due to the higher probability of short circuits.

On the other hand, the probability of short circuits is higher in gas-assisted machining due to the smaller width of the inter-electrode gap (Fig. 7.13b). Increased performance is achieved by using ultrasonic vibration to assist the process, which is performed by the workpiece in the described studies [19].

This method has been in the research stage for many years. Its unquestionable advantage is the elimination of the problem of the adverse environmental effects of oil dielectric fluid and its decomposition products during discharges.

7.3.2.1 Gaseous Media-Assisted ECM Process

Electrochemical jet machining has been of interest to technologists for many years [20, 21]. The tool is an electrolyte jet flowing out of a micro-nozzle at high speed. The diameter of the jet can be small and concentrated on a small area ($d < 0.5$ mm). An electric field is generated between the nozzle and the workpiece. An electric current flows through the electrolyte and the material is removed by electrochemical dissolution (Fig. 7.15a). It should be noted that by adequately moving the workpiece or the electrolyte jet, small 3D surfaces can be shaped (Fig. 7.16).

The advantage of electrochemical machining is the fact that the direct factor responsible for the localisation of the dissolution process and material removal is a flexible tool, i.e. the high-pressure electrolyte jet. Material removal is a hybrid process because it combines electrochemical reactions with the mechanical impact

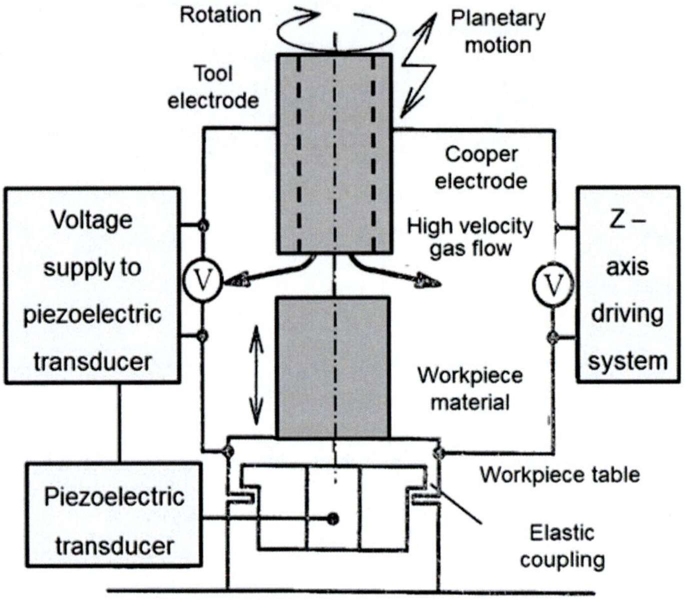

Fig. 7.14 Scheme of the EDM process in a gaseous atmosphere with ultrasonic assistance [19]

of the jet. The electrochemical jet process can also be assisted by the simultaneous use of an electrolyte jet and a gas jet (Fig. 7.15b) [2, 22] or by increasing the mechanical action of the electrolyte jet by adding abrasive particles to the electrolyte [23].

As mentioned earlier, the main difficulty in the application of classical electrochemical machining in the shaping of small-sized components is the localisation of the dissolution process on a small area. This is related, among other things, to the difficulties of making an electrode of small dimensions, supplying the electrolyte to the

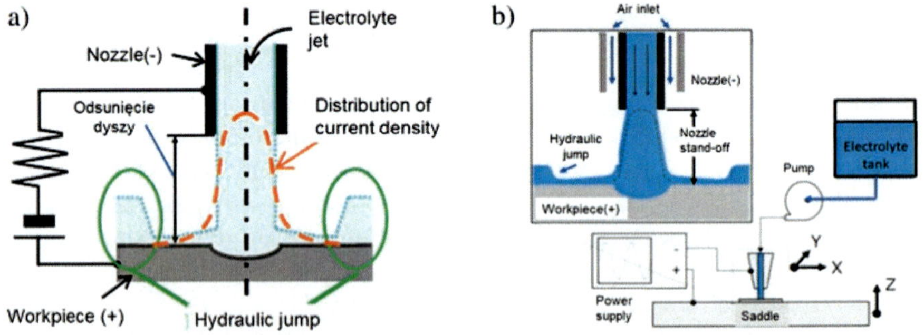

Fig. 7.15 Principle of electrochemical jet machining **a** and additional shield of pressured air **b** [21, 24]

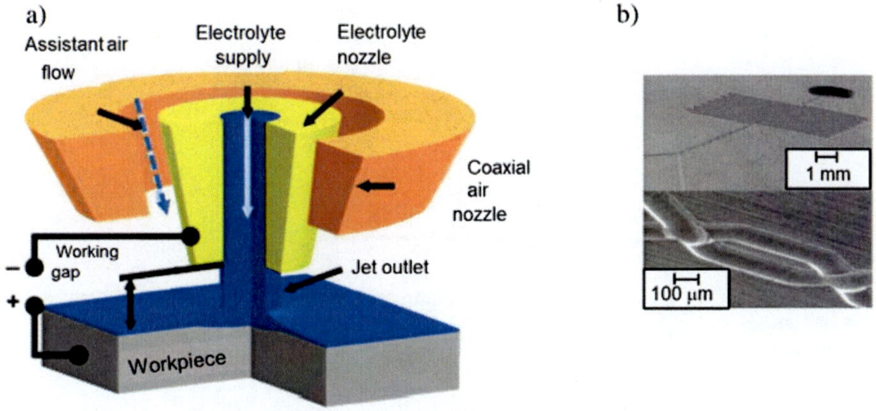

Fig. 7.16 Scheme of elastic production system for utilization of Jet ECM process **a** and example of fabricated microstructure from stainless steel **b** [2, 22]

machining zone and removing the products of the dissolution (debris). The problem is solved by using an electrolyte jet or an electrolyte jet with abrasive particles [23].

Electrochemical slurry jet machining (ESJM) is a combination of abrasive slurry jet machining (ASJM) and electrochemical jet machining (ECJM) [23]. The working fluid consists of Al_20_3 grains and a mixture of $NaNO_3$ and NaCl electrolytes (Fig. 7.17). In the tests, a constant voltage of 30 V was applied between the nozzle and the workpiece. Grooves were made in Stellite 12 optionally using ASJM, ECJM and ESJM. The results show that the removal rate of the ESJM process increases significantly compared to separate use of ASJM and ECJM due to the synergy effect. The contribution of ASJM increases along with increasing kinetic energy of Al_20_3 particles (pressure increase) and abrasive particle concentration. What seems to be

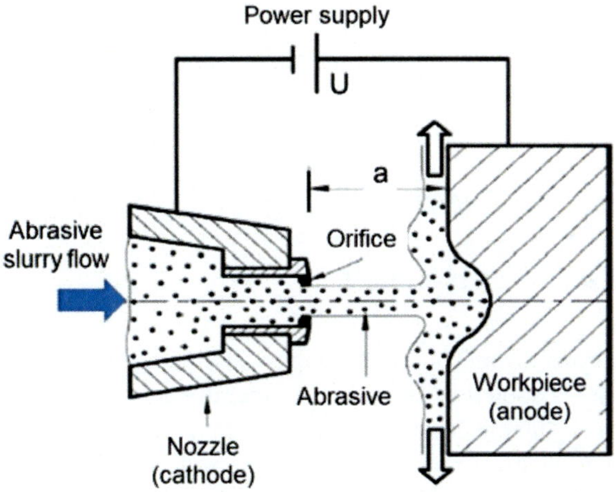

Fig. 7.17 Scheme of electrochemical-assisted slurry jet machining(ESJM). Workpiece material: Stellite 12, abrasive grains—Al_2O_3 [23]

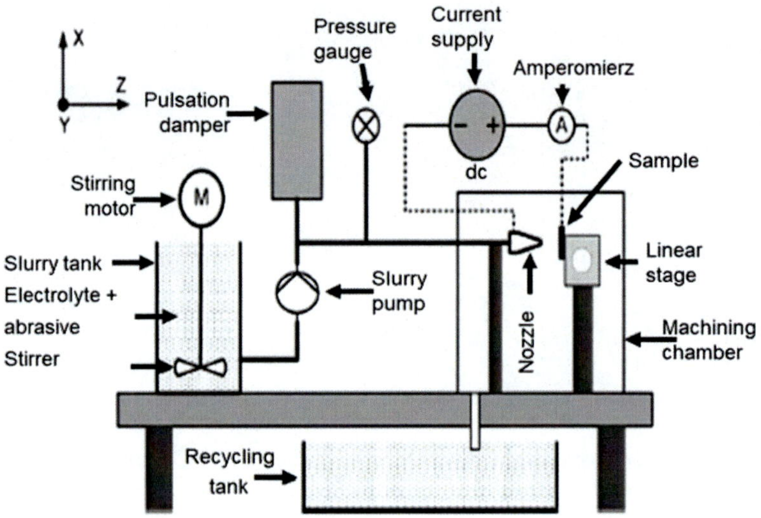

Fig. 7.18 Scheme of set-up for investigating ASJM, ECJM and ESJM processes [23]

obvious, the surface roughness decreases with an increase of the contribution of abrasive erosion.

In the study of the ASJM, ECJM and ESJM processes the same conditions were applied (Fig. 7.18): jet diameter −180 μm, jet pressure −1.5–3.0 MPa, jet temperature −25 °C, nozzle distance from the workpiece material −2.5 mm, type of abrasive grain and its concentration (in ECJM–no abrasive grains were used), concentration of abrasive grains 0.125–0.5%, voltage U = 30 V (not used in ASJM), electrolyte mixture NaCl + $NaNO_3$ with concentration of 7.5% + 7.7% (not used in ASJM). Micro-shaping of surface structures in special materials by ASJM, ECJM and ESJM methods is dynamically developed. The presented research results indicate that they offer great technological opportunities for the processing of special materials with high hardness, e.g. Stellites. From the point of view of material removal rate and surface quality, favourable results are given by the ESJM method [23].

7.4 Cryogenic Subtractive Processes

7.4.1 Introduction

Cryogenic machining/cutting is classified as a group of processes assisted by gaseous media cooled to cryogenic temperatures [4]. It is generally accepted that the onset of cryogenics corresponds to a temperature of −150 °C (123 K), although CASA (Cryogenic American Society of America) defines it below 120 K (−53 °C) and NIST below −180 °C [8]. This corresponds to a temperature below the boiling point of many gases–helium, hydrogen, oxygen, nitrogen and normal air. The cryogenic media of technological importance are liquid nitrogen (LN_2) at −186 °C, helium (LHe) and cooled carbon dioxide (LCO_2) at approximately (−50 °C)–(−78 °C). This means that contrary to the accepted definitions in technological practice, cryogenics refers to cooling media with sub-zero temperatures, i.e. below 0 °C. Figure 7.19 shows phase transformations for carbon dioxide (a) and nitrogen (b). Liquid nitrogen is stored under vacuum in a Dewar tank at a pressure of 4 bar (0.4 MPa) and delivered to the cutting zone at −195.8 °C (77.35 K). CO_2 on the other hand, is supplied at ambient temperature of 20 °C under pressure of 57 bar (5.5 MPa). Due to evaporation and sublimation at the nozzle outlet, it cools down to the boiling point, i.e. −78.5 °C (194.15 K). Note that the critical point is 74 bar/31 °C for CO_2 and 34 bar/−147 °C for N_2.

When using cryogenic media, the time during which the jet reach the cutting zone is important. In the case of liquid nitrogen LN_2 (Fig. 7.19c), it is significantly longer than for CO_2. The CO_2 jet flowing out of the nozzle reaches the minimum temperature of −50 °C after approximately 60 s. Even when liquid nitrogen penetrates the cutting

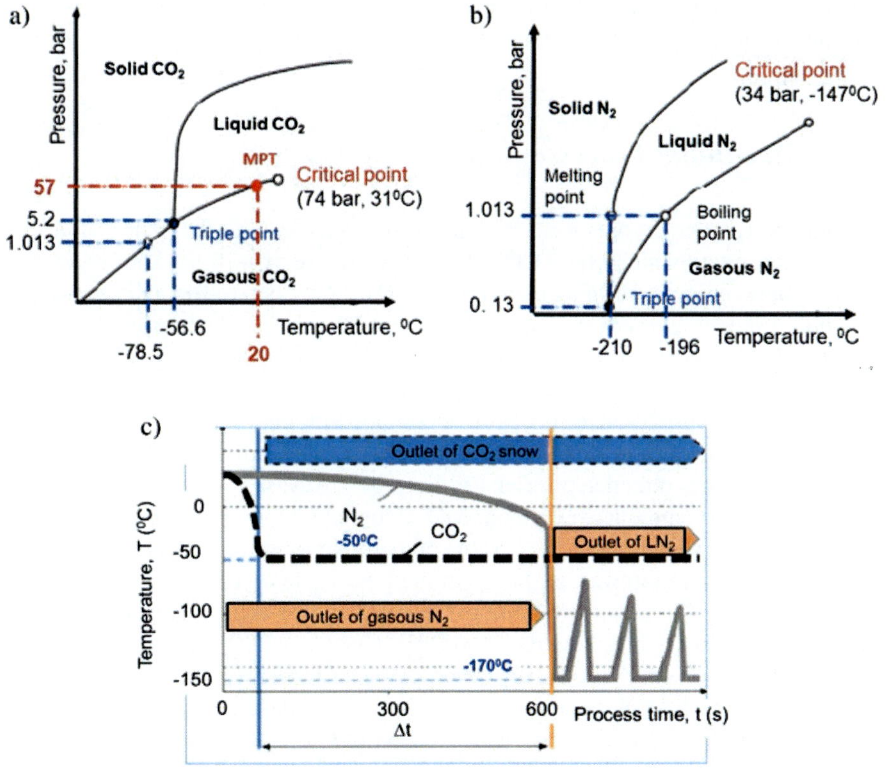

Fig. 7.19 Phase transitions for CO_2 **a** and nitrogen N_2 **b**; differences in time of medium supply to treatment zone **c** [11, 25, 26]

Table 7.3 Specification of potential advantages and disadvantages of cryogenic machining [27]

Criteria	Advantages	Disadvantages
Costs	Higher cutting speed/longer tool life	Additional cost of cooling medium
Part quality	Better surface quality	Lower dimensional accuracy
Environment/society	Clean manufacturing environment	Defrosting/partial corrosion

zone and reaches the temperature of −170 °C after 600 s, pulsations still occur due to the presence of the gaseous phase.

Hybrid cooling strategies are also currently developed (Table 7.1) in which the cryogenic effect is assisted by lubrication with an aerosol oil suspension (MQL) with an LN_2 or LCO_2 jet. Table 7.3 summarises the advantages and disadvantages of cryogenic machining. Unquestionable are the environmental benefits and the social aspects, e.g. the most important one concerning the absence of toxic effects on staff's health. On the other hand, the economic side of the process is questioned due to the high price of cryogenic media (a high flow rate is required to cool sufficiently the tool, as the effect spreads to the workpiece and machine tool components) and the additional costs associated with their storage and transport.

7.4.2 *Machining Processes with Cooled CO_2*

As mentioned earlier, cryogenic machining with cooled CO_2 uses the Joule–Thomson effect, according to which liquid CO_2 at the temperature of −78.5 °C boils (foams) at atmospheric pressure forming a mixture of CO_2 snow foam (60%) and gas (40%). In its liquid state it is stored at a pressure of about 50 bar. Cooling results from rapid expansion by evaporation (sublimation) when the liquid CO_2 flows out of the nozzle. For this reason, the position of the nozzle relative to the zone to be cooled is very important. An important factor in effective cooling with the CO_2 jet is the flow rate, the increase of which results in increased heat dissipation from the cutting zone [28].

The wear curves obtained after end-milling of Ti-6Al-4 V titanium alloy using different cooling techniques, i.e. flood, MQL, cooled CO_2 and hybrid (CO_2+MQL and CO_2+air) are shown in Fig. 7.20a. In comparative wear tests, a constant cutting velocity $v_c = 100$ m/min was used for a limited wear rate of $VB_B = 0.2$ mm. It can be seen from Fig. 7.20a that a tool life of T = 30 min was achieved using emulsion and pressure flood cooling (curves #2 and #1). Cryogenic CO_2 cooling (curve #5) did not reduce tool wear in this case, even compared to dry cutting. On the other hand, hybrid CO_2 + MQL cooling (curve #6) proved to be more effective, but for a shorter tool life of about 15 min [29]. Therefore, in Fig. 7.20b, the tool life periods were estimated for different cutting speeds. It can be concluded that the addition of CO_2 to the aerosol increases the tool life by 30% on average, but for lower cutting speeds (for $v_c = 100$ m/min) the MQL method is more effective.

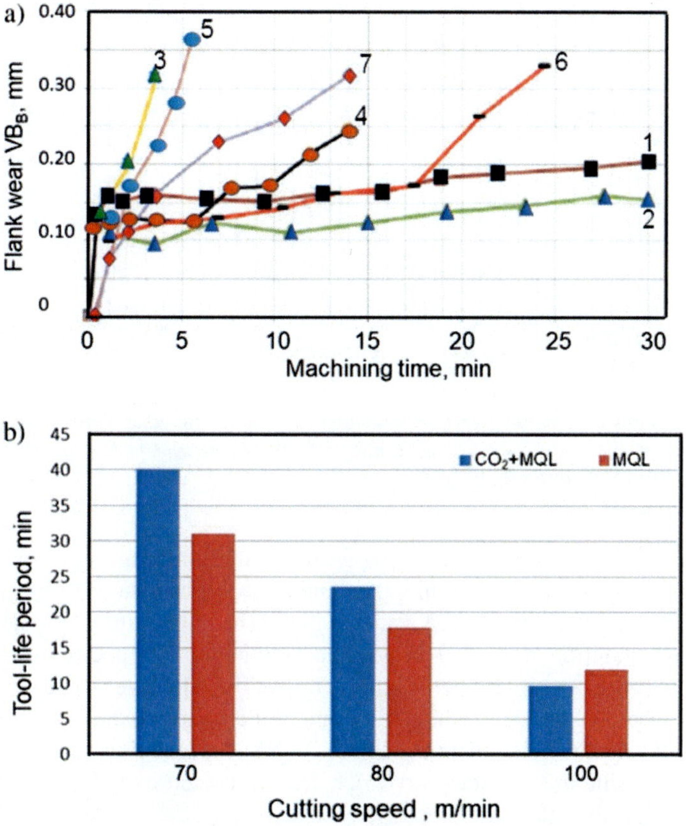

Fig. 7.20 Comparison of effects of different cooling/lubricating systems in terms of tool wear for the cutting speed of $v_c = 100$ m/min; symbols of cooling systems: 1—HP with medium pressure, 2—flood, 3—dry, 4—MQL, 5—CO_2, 6—CO_2+MQL, 7—CO_2+air **a** and tool life **b** [29]

7.4.3 Cryogenic Machining Processes with Liquid Nitrogen (LN2)

The documented effects of LN_2 on the cutting process include reduced plastic deformation, reduced friction, reduced abrasive wear on the flank face and lower tendency of crater formation on the rake face. These findings have been demonstrated in the machining of titanium alloys, superalloys (e.g. Inconel 718 alloy), shape memory alloys, magnesium alloys, Co-Cr–Mo bio-alloys, composites based on metal matrix, hard steels, cast irons and medical materials [1].

The current state of application of cryogenic cooling technology in relation to material grade and type of operation indicates that the turning of hard steels is predominant, but there is a noticeable lack of drilling due to the difficulty of adjusting nozzles to supply the cryogenic medium into the machining zone. The machining

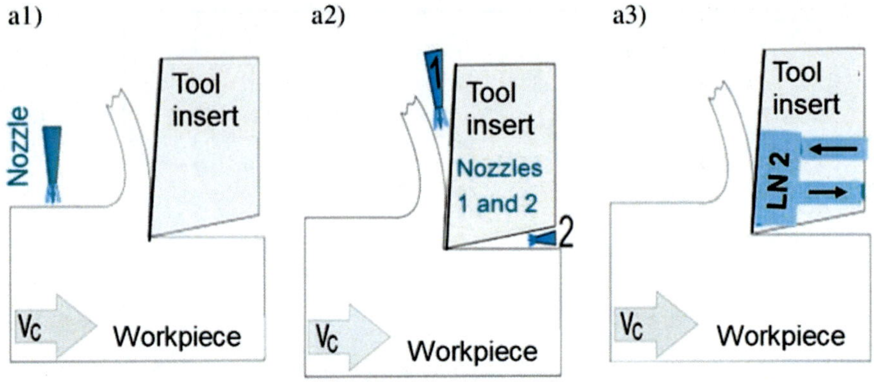

Fig. 7.21 Methods of supplying cryogenic medium to the cutting zone: (a1) through the outer nozzle to the machined surface, (a2) using one or two outer nozzles to the contact zones, (a3) through channels inside the tool [26, 30]

of ductile materials and polymers (usually under micro-machining conditions), on the other hand, involves transforming the material to a brittle state. There is even the concept of cooling the formed chip, which will then be more brittle. Very demanding materials are titanium and nickel alloys, and only about 13% and 6.5% of studies, respectively, concern their cryogenic milling operations. However, there are real indications for improving their machinability and reducing the cost of their machining [30]. It should also be noted that 68% of the cutting tool materials used in cryogenic machining are sintered carbides, of which 1/3 are coated carbides. This is easily explained due to the increase in tool life obtained.

Methods of supplying liquid nitrogen to the machining zone are shown in Fig. 7.21. Generally, cooling of the cutting edge is used (Figs. 7.21a2 and a3), but it is also possible to preliminarily cool the outer surface of the workpiece (Fig. 7.21a1). It is also possible to cool the entire workpiece thoroughly in a cryogenic chamber and quickly transport it to the machine tool. In the case of cutting tool cooling (which is the most effective) the cooling medium in the form of a jet of liquid gas is supplied externally through one (between the chip and the contact surface) or two nozzles (additionally from the side of the flank face) and internally through special channels. LN_2 absorbs the heat generated in the tool—chip contact area, evaporates and forms a gas–liquid (GN_2) shield which acts as a lubricating film.

For example, in the cryogenic machining of titanium alloys the tool life can increase tenfold and the material removal rate twofold. On the one hand, the cutting force components increase as a result of the hardening of the workpiece material, but the tool life also increases due to the intensive heat dissipation through the cryogenic medium.

Figure 7.22 shows a comparison of the intensity of wear evolution on the cutter edges in cryogenic milling of Ti-6Al-4 V titanium alloy using CO_2 and LN2. Less and more uniform wear can be seen when liquid nitrogen is supplied to the contact zone instead of CO_2. This effect is explained by the formation of thin liquid/gas

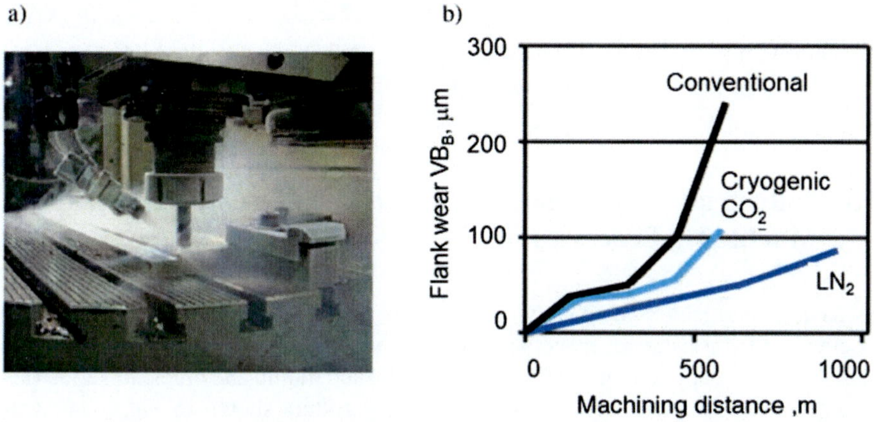

Fig. 7.22 Cryogenic milling of titanium alloy Ti-6Al-4 V (D_{cut} = 12 mm, a_e = 8 mm, a_p = 5 mm, v_c = 100 m/min, f_z = 0.03 mm) **a** and comparison of wear evolution for CO_2 and LN2 cooling methods **b** [6]

layers between the sliding surfaces of the tool and the chip, which reduces friction (μ values down to 0.3–0.4) [30]. Cryogenic machining is an effective way to improve the machinability of materials with extremely poor machinability, such as intermetallics (titanium aluminides) used increasingly for gas turbine blades.

Figure 7.23 shows the effects of different cooling strategies (including LN_2) on

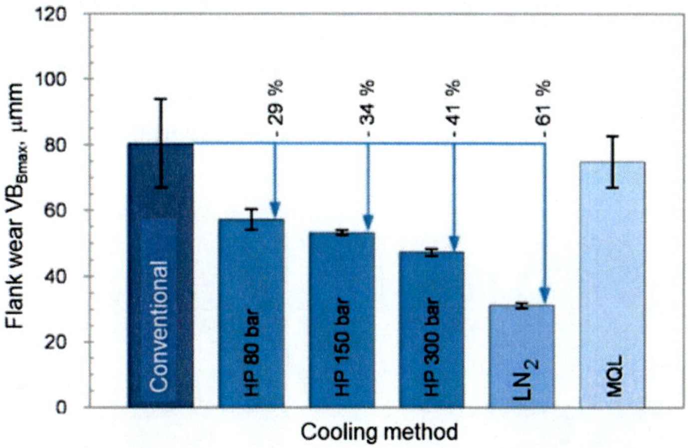

Fig. 7.23 Comparison of tool wear in the machining of γ-TiAlN intermetallic alloy using HW-P10 carbide tool for different cooling methods; machining parameters: $v_c = 80$ m/min, $f = 0.1$ mm/rev, $a_p = 0.25$ mm [6, 8]

tool wear (keeping VB_{Bmax} index) in the cutting of intermetallic γ-TiAlN with a HW-P10 carbide tool. It can be clearly seen that the LN2 supply resulted in a reduction in tool wear of about 60% compared to gravitation cooling.

7.5 Industrial Applications of Liquid and Gaseous-Assisted Machining Processes

As frequently mentioned in previous chapters, hybrid processes with liquid and gaseous media assistance, especially the HPC and cryogenic processes have so far found numerous applications in the aerospace, die and mould and precision industries [1, 2]. Universal tools, such as *cryo-tec* milling cutters shown in Fig. 7.24, with optional supply of various cooling/lubricating media, i.e. air, emulsion, aerosol, CO_2 are increasingly used for machining operations.

An example of a practical implementation of an HPC process based on *Beyond Blast PCT (Precision Coolant Technology)* is shown in Fig. 7.25a. In this cooling solution, CF is supplied through special channels in the cutting insert at a low pressure of about 7 bar (Fig. 7.25b), which, however, has a similar effect as the high pressure of about 70 bar in the external CF supply. This solution has resulted in a 70–100% increase in tool life in the machining of titanium alloys [8].

In the case of cryogenic machining with liquid nitrogen, commercial LN2 distribution systems are already in use for CNC machining centres, as shown in Fig. 7.26, equipped with appropriate installations and tools. Turning tools use cutting inserts with internal channels supplying the LN2 jet (Fig. 7.27b), while in milling tools the LN2 jet flows through the spindle and internally through the channels (Fig. 7.27c). The different ways of supplying LN2 to the cutting zone are shown in Fig. 7.21.

Cryogenic machining is also used to machining ceramics with PCBN tools, or polymeric materials such as the thermoplastic PMMA-polymethyl methacrylate used

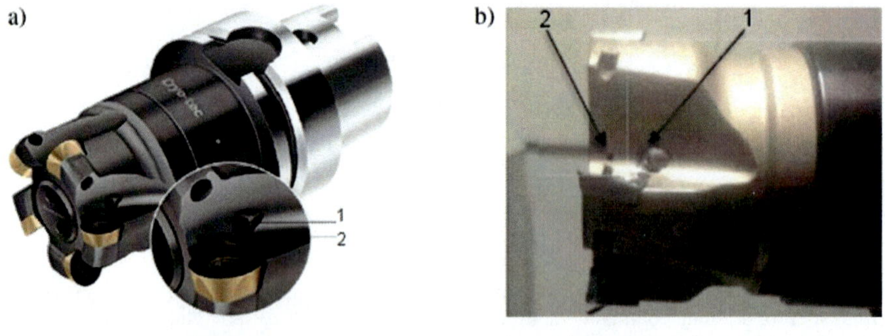

Fig. 7.24 Examples of milling cutters adopted to optional supply of different cooling media for cryogenic machining (*cryo-tec* type) through internal channels (1-air, emulsion and aerosol, 2-cryogenic CO_2): copying cutter **a** [30], end-milling cutter (1-MQL and/or air, 2-CO_2) **b** [29] (www.kennamatal.com)

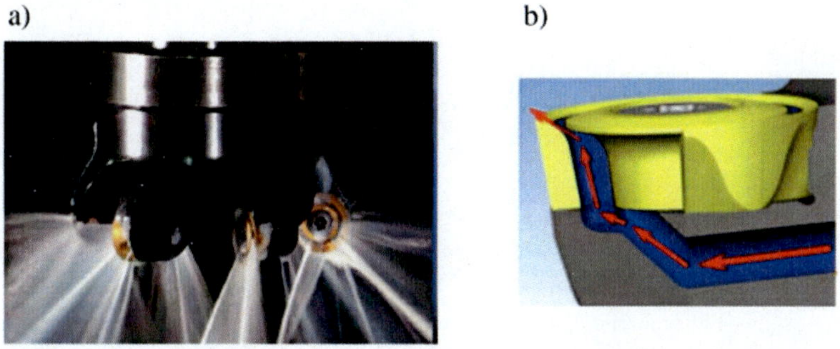

Fig. 7.25 View of a milling cutter head of *Beyond Blast PCT (Precision Cooling Technology)* type **a** with delivering CF through internal channels in cutting insert (b) [5] (www.kennamatal.com)

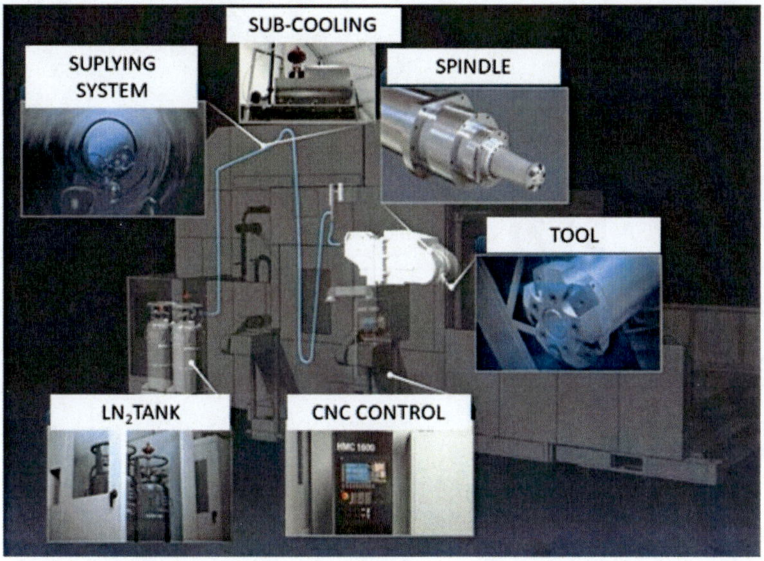

Fig. 7.26 Industrial system for liquid LN2 distribution to hybrid machine tools (adapted to cryogenic machining) [5, 31]

for implanted intraocular lenses (IOLs). In these cases, a definite improvement in surface smoothness is required. The effects of cryogenic turning of PMMA polymer are shown in Fig. 7.28a, b. As the material after freezing at −40 °C (Fig. 7.28c) changes into a glassy state, a mirror-like hydrophilic surface is obtained (Fig. 7.28b) with much lower roughness (the surface roughness parameter *Rq* is 40% lower).

Figure 7.29 summarises the optimum application of the various strategies for media-assisted processes in the machining of heat-resistant alloys, which are classified as materials with very poor machinability [1]. From the proposals for the application of various cooling and lubrication strategies, it is clear that the application of HSC (*High Speed Cutting*) and HPC (*High Performance Cutting*) techniques

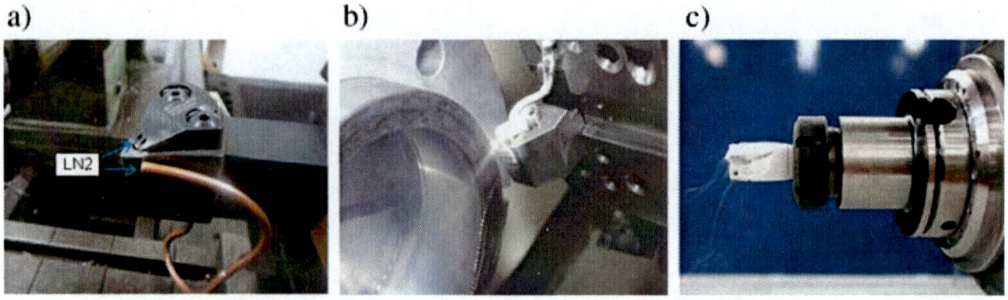

Fig. 7.27 Methods of supplying liquid LN2 to the machining zone: through the internal channels in cutting tool insert and additional outer nozzle **a**, outer ICEFLY system **b** [27], **c** through the spindle of milling machine for cryogenic MQL method [29]

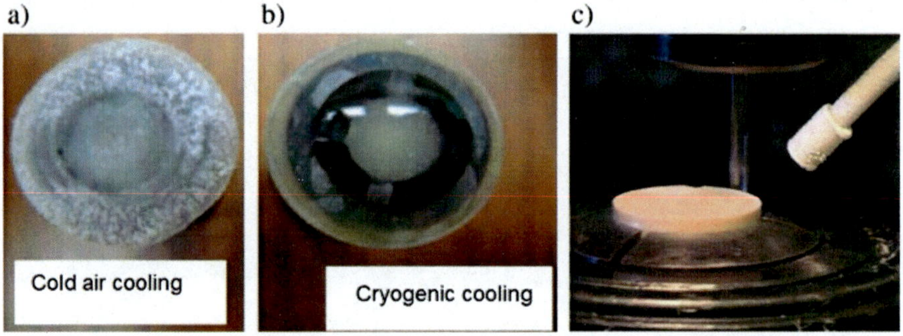

Fig. 7.28 Images of implanted intraocular lens made from thermoplastic biopolymeric material (PMMA) after conventional machining with cold air cooling **a** and after cryogenic machining using ICEFLY system shown in Fig. 7.27b) **b** [32], **c** freezing of the machined element

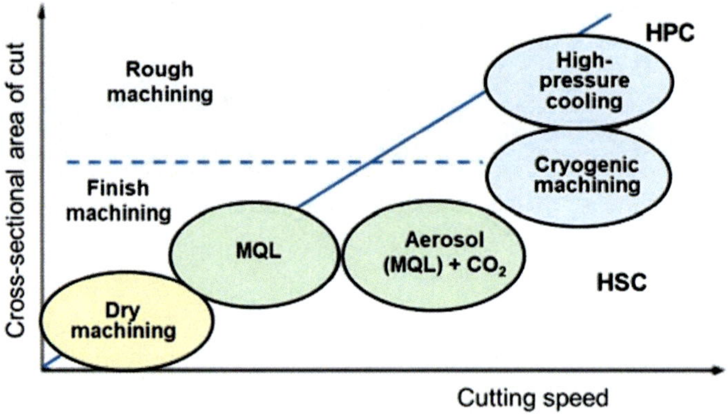

Fig. 7.29 Application range of various cooling/lubrication strategies in the machining of heat resistant alloys [9, 25]

is only possible with high-pressure liquid assistance and cryogenic assistance. In finishing, a technique combining minimum lubrication MQL (in the original ADL-Aerosol *Dry Lubrication*) with additional cooling LCO_2 is acceptable [9]. Hybrid ($MQL+CO_2$) machining has proven to be acceptable in the machining of Inconel 718 alloy, among others.

References

1. Grzesik W (2018) Advanced machining processes of metallic materials. Elsevier, Amsterdam
2. Lauwers B, Klocke F, Klink A, Tekkaya AE, Neugebauer R, McIntosh D (2014) Hybrid processes in manufacturing. Ann CIRP Manuf Technol 63(2):561–583
3. Luo X, Qin Y (2018) Hybrid machining. Academic Press (Esevier), London
4. Abdelrazek A, Choudhury I, Nukman Y, Kazi S (2020) Metal cutting lubricants and cutting tools: a review on the performance improvement and sustainability assessment. Int J Adv Manuf Technol 106:4221–4245
5. Internet: High pressure coolant machining. www.sandvik.coromant.com; Precision cooling technology. www.kennametal.com
6. Courbon C, Kramar D, Krajnik P, Pusavec F, Rech J, Kopac J (2009) Investigation of machining performance in high-pressure jet assisted turning of Inconel 718: an experimental study. Int J Mach Tools Manuf 49:1114–1125
7. Shokrani A, Dhokia V, Munoz-Escalona P, Newman ST (2013) State-of—the—art cryogenic machining and processing. Int. J Comp Integr Manuf 26:616–648
8. Jawahir IS, Attia H, Biermann D, Duflou J, Klocke F, Meyer D, Newman ST, Pusavec F, Putz M, Rech J, Schulze V, Umbrello D (2016) Cryogenic manufacturing processes. Ann CIRP Manuf Technol 65(2):713–736
9. Busch K, Hochmuth C, Pause B, Stoll A, Wertheim R (2016) Investigation of cooling and lubrication strategies for machining high-temperature alloys. Proc CIRP 41:835–840
10. Ayed Y, Germain G (2018) High-pressure water-jet-assisted machining of Ti555-3 titanium alloy: investigation of tool wear mechanisms. Int J Adv Manuf Technol 96:845–856
11. Sterle L, Grguras D, Kern M, Pusavec F (2019) Sustainability assessment of advanced machining technologies. J Mech Eng 65:671–679
12. Braham-Bouchnak T, Germain G, Robert P, Lebrun JL (2010) High pressure water jet assisted machining of duplex steel: machinability and tool life. Int J Mater Form 3:507–510
13. Prihandana G, Mahardika M, Sriani T (2020) Micromachining in powder-mixed micro electrical discharge machining. Appl Sci 10:3795
14. Molinetti A, Amorim FL, Soares PC Jr, Czelusniak T (2016) Surface modification of AISI H13 tool steel with silicon or manganese powders mixed to the dielectric in electrical discharge machining process. Int J Adv Manuf Technol 83:1057–1068
15. Zhu Z, Dhokia VG, Nassehi A, Newman ST (2013) A review of hybrid manufacturing processes-state of the art and future perspectives. J Comp Integr Manuf 26(7):596–615
16. Singh AK, Kumar S, Singh VP (2015) Effect of the addition of conductive powder in dielectric on the surface properties of superalloy Super Co 605 by EDM process. Int J Adv Manuf Technol 77:99–106
17. Unune D, Mali H (2015) Current status and applications of hybrid micro-machining processes: a review. Proc IMEng—J Eng Manuf 229(10): 1681–1693
18. Kunieda M, Yoshida M (1997) Electrical discharge machining in gas. Ann CIRP 46(1):143–146
19. Yoshida M, Kunieda M, Kaneko Y (1998) Improvement of material rate of dry EDM using piezoelectric actuator coupled with servo-feed mechanism. Processes 14th international conference CAPE CAPE'98 Tokyo pp 283–288

20. Kozak J, Rajurkar KP, Balkrishna R (1996) Study of electrochemical jet machining process. Trans ASME J Eng Ind 118(4):490–498
21. Masuzawa T, Takawashi T (1998) Recent trends in EDM/ECM technologies in Japan. Processes international symposium electrical machines ISEM XII Aachen Germany pp 1–15
22. Hackert-Oschaetzchen M, Meichsner G, Zinecker M, Martin A, Schubert A (2012) Micro machining with continuous electrolytic free jet. Prec Eng 36:612–619
23. Liu Z, Noruaei H, Papini M, Spelt JK (2014) Abrasive enhanced electrochemical slurry jet micro–machining: comparative experiments and synergistic effects. J Mater Proc Technol 214:1886–1894
24. Walker C, Kamps TJ, Lam JW, Mitchell-Smith J, Clare AT (2017) Tribological behaviour of an electrochemical jet machined textured Al-Si automotive cylinder liner material. Wear 366–367:1611–1621
25. Blau P, Busch K, Dix M, Hochmuth C, Stoll A, Wertheim R (2015) Flushing strategies for high performance, efficient and environmentally friendly cutting. Proc CIRP 26:361–366
26. Pusavec F, Grguras A, Koch M, Krajnik P (2019) Cooling capability of liquid nitrogen and carbon dioxide in cryogenic milling. Ann CIRP Manuf Technol 68(1):73–76
27. Lu T, Dillon OW, Jawahir IH (2013) A thermal analysis framework for cryogenic machining and its contribution to product and process sustainability. Proceedings 11th global conference on sustainable manufacturing—innovative solutions Berlin pp 262–267
28. Sadik MI, Isakson S, Malakizadi A, Nyborg L (2016) Influence of coolant flow rate on tool life and wear development in cryogenic and wet milling of Ti-6Al-4V. Proc CIRP 46:91–94
29. Tapoglou N, Lopez MA, Cook I, Taylor ChM (2017) Investigation of the influence of CO_2 coolant application on tool wear. Proc CIRP 63:745–749
30. Stefansson T (2014) Application of cryogenic coolants in machining processes. KTH, Stockholm
31. M'Saoubi R, Axinte D, Leung Soo S, Nobel Ch, Attia H, Kappmeyer G, Engin S, Wei-Ming S (2015) High performance cutting of advanced aerospace alloys and composite materials. Ann CIRP Manuf Technol 64(2):557–580
32. Knopf J, Grimm L, Ghosh R, Gibson D (2013) Cryogenic machining of PEEK. Proceedings 1st PEEK conference Philadelphia

Chapter 8
Magnetic and Electric Field-Assisted Machining Processes

8.1 Introduction

During a long time the development of unconventional material removal processes in relation to traditionally used metallic and no-metallic materials substantially progressed but recently a variety of new advanced materials, which allow to improve part functionality, and corresponding manufacturing methods have been dynamically investigated and utilized in many industry branches. It is obviously known that the application of new materials generates a series of serious technological problems including the dimensional accuracy of machined parts. This is especially related to the dynamic growth of the manufacturing segment of miniature-electro-mechanical systems in both micro and nanoscale (MEMS and NEMS). It is well known that these systems include various micro and nano-actuators (see for instance Fig. 3.5 in Sect. 3.3). As a result, traditional machining methods are improved and new machining processes are developed in order to achieve more efficient and more accurate material removal processes. Among these machining activities some leading trends concern the concurrent application of various processes or wider application of hybrid processes with a predominant participation of non-traditional EDM and ECM processes. A general characterization of hybrid machining processes is presented in Sect. 1.3 but this chapter is devoted to magnetic field and external electric field assisted machining processes covering finish abrasive processes, i.e. polishing and lapping operations as well as non-traditional EDM and ECM processes. The main technological philosophy is that machining parameters should be selected to guarantee the demanded part accuracy and surface integrity along with minimum wear of the tool-electrode and as low as possible machining cost. Additional technical problem which appears during process selection is the part dimension and associated dimension of the machining zone. It was documented that during machining of macro-parts the influence of the assisted magnetic field is marginal but, in sharp contrast, it becomes decisive when machining micro-parts with a maximum dimension less than 1 mm [1–3]. The main reason is that erosion products are more easily

W. Grzesik and A. Ruszaj, *Hybrid Manufacturing Processes*, Springer Series in Advanced Manufacturing, https://doi.org/10.1007/978-3-030-77107-2_8

evacuated from the discharge zone, which improves process efficiency, part accuracy and surface finish.

Hybrid ECM processes can be divided, taking into account their participation in the material removal, into two groups, i.e. assisted and pure (combining) processes. In the case of hybrid ECM processes the main material removal mechanism, i.e. electrochemical reactions resulting in electrochemical dissolution, is assisted by external energy source, as for instance by vibrations applied on the working electrode, the workpiece or the working fluid (electrolyte), and laser heating of the removed material layer. On the other hand, for pure ECM processes material removal results from the integration of various mechanisms, as for instance in simultaneous EDM/ECM (SEDCM) process or in ECM-assisted grinding (ECAG). In these processes the effect of process synergy occurs but all componential processes take part separately in material removal and the generation of the surface integrity.

It is proposed in Chap. 6 that, independently of the machining zone dimension, both hybrid EDM and ECM processes can be divided into two main groups, namely: assisted and pure (combining) processes. Among assisted hybrid non-traditional material removal processes the following main processes are applied [1–3]:

1. Magnetic field-assisted EDM (MFAEDM) process,
2. Magnetic field-assisted ECM (MFAECM) process,
3. Electric field-assisted EDM process.

In the future, new assisted and pure hybrid processes can be developed and consequently their practical applications are also possible [1–3].

8.2 Magnetic Field-Assisted Finishing Processes

Among the micro-/nano machining (MNM) processes a special group of magnetic abrasive finishing processes is specified [3, 4]. It includes the following processes: Magnetic Abrasive Finishing (MAF) also known as Magnetic Field-Assisted Abrasive Flow Machining (MAAFM), Magnetorheological Finishing, Magnetorheological Abrasive Flow Finishing, and Magnetic Float Polishing. The general rule is that these processes use the magnetic abrasive particles (MAPs), which are composed of ferromagnetic materials such as iron particles and non-magnetic abrasive powders, whose properties can be controlled externally with the help of magnetic field. Typically, silicon carbide (SiC) and alumina oxide (Al_2O_3) particles of different size are used. The abrasive and ferromagnetic particles are mixed with a small amount of metal working fluid (oil, kerosene). The electromagnetic behaviour of MAPs can be either static or dynamic depending on the excitation of the electromagnet. When DC power is used, the process is called *direct current abrasive finishing* (DCAF) and the magnetic brush so formed is called a *static-flexible magnetic abrasive brush* (S-FAMB). This permits to control the forces acting on an abrasive particle and, as a result, the amount of material removed is also controlled. This class of processes is capable of producing surface roughness value of 8 nm or lower. Using better force control and still finer abrasive particles, some of these processes may result in the sub-nanometer surface roughness value on the finished part. In addition, the MAF process can remove surface defects produced in micro-grinding, electro-discharge micromachining and laser beam micromachining. Further hybridization of the MAF process involves ultrasonic-assisted magnetic abrasive finishing (UAMAF) and electrolytic abrasive finishing process (EMAF) [3].

The schemes of the MAF process applied to cylindrical and plane surfaces are presented in Figs. 8.1 and 8.2 respectively. During the process the workpiece is kept between the N-pole and S-pole of the magnet and the working gap between the workpiece surface and the magnet is filled with a magnetic abrasive. In consequence, abrasive particles joined magnetically between magnetic poles N and S act as a multipoint abrasive tool, i.e. a *flexible magnetic abrasive brush* (FAMB), and a workpiece surface inserted into so formed processing zone is finished. The external cylindrical MAF process (Fig. 8.1a) is performed when the cylindrical workpiece starts to rotate at the same time as it vibrates along its axis. In internal MAF process the process kinematics includes the rotations of the magnetic pole and the workpiece (tube). During the process MAPs introduced into the workpiece are attracted by the magnetic field to bear down on the inner workpiece surface (Fig. 8.1b).

A MAF schematic diagram related to finishing of flat surfaces is shown in Fig. 8.2. The process principle is a similar type as shown in Fig. 8.1. A round, flat-faced electromagnet with a ferromagnetic centre pole and an electrical coil surrounded by an outer shell is used. When this system is activated, a strong magnetic field in the centre pole generates radially oriented magnetic force which, in the combination with pole rotation, develops abrasion pressure for finishing of the flat surface. It should be noted that the self-adaptive AMB is generated. The enlarged finishing zone shown in Fig. 8.3 illustrates the distribution of forces acting on the work surface during material removal process and subsequent stages of material removal. The normal/radial force (F_n) generated by the magnetic field causes that abrasives penetrate inside the workpiece materials whereas tangential/axial force (F_t) generated by rotation of the magnetic abrasive brush (i.e. N-pole) results in the removal of formed micro-chips.

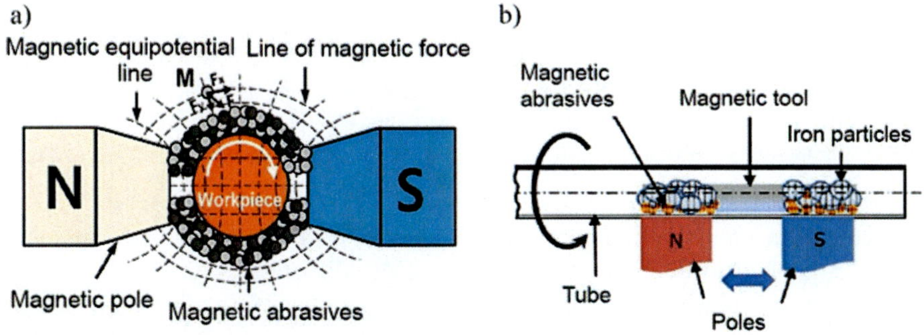

Fig. 8.1 Scheme of magnetic field- assisted finishing of cylindrical external (**a**) and internal (**b**) workpiece [4]

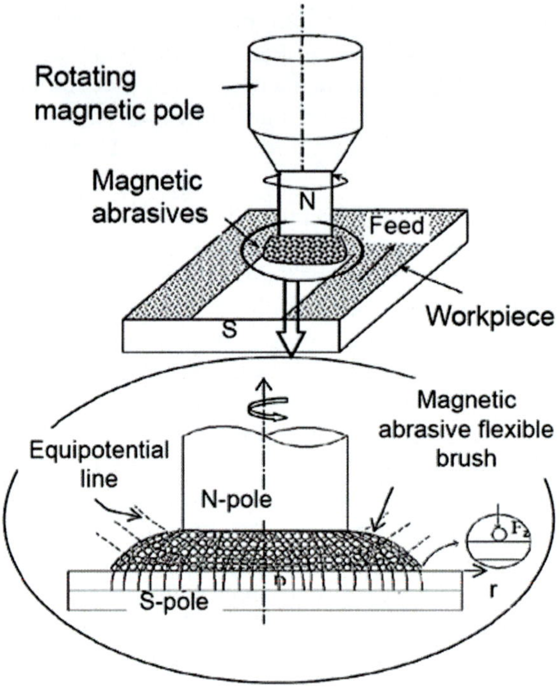

Fig. 8.2 Scheme of magnetic field- assisted finishing of a plane workpiece [1, 4, 5] (www.home.iitk.com.in)

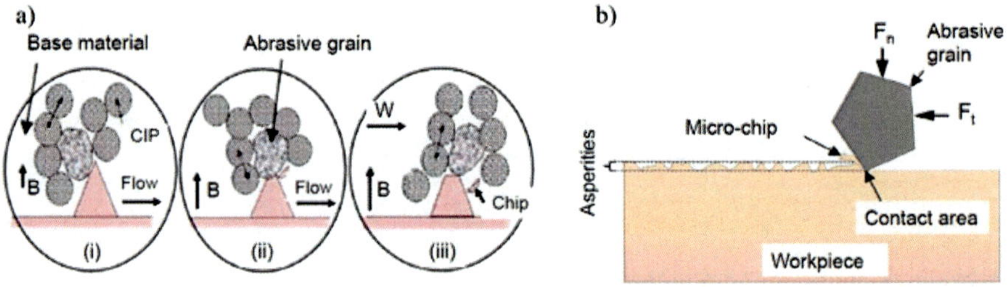

Fig. 8.3 Stages in removing irregularities on a flat surface (**a**) and forces acting on magnetic abrasive particle [4, 5]. Symbols: B—bunch of iron/abrasive particles, W—rotation effect of magnetic brush, CIP—carbonyl iron-abrasive particle

Figure 8.4a presents an improved concept of AFM process called rotational magneto-rheological abrasive flow finishing (MRAFF) process. It is basically a combination of abrasive flow machining (AFM) and magnetorheological finishing (MRF). This process has been developed for nano-finishing of internal and external complex parts' geometries for a wide range of industrial applications [7]. During finishing a reciprocating and rotational motions to the abrasive media are utilized by

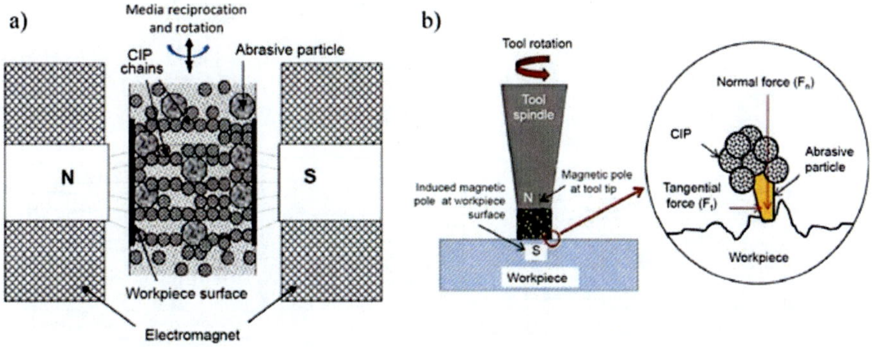

Fig. 8.4 Scheme of rotational magneto-rheological abrasive flow finishing process (MRAFF) (**a**) and ball end magnetorheological finishing (BEMRF) [6, 7]

the hydraulic cylinder and rotated magnetic field. It imparts better process performance in comparison to AFM process due to in-process control of abrading medium's rheological behaviour through magnetic field. Because the MRF process is incapable to finish 3D intricate shaped surfaces, a new precision nanofinishing process of 3D surfaces using a ball-ended tool termed ball end MR finishing (BEMRF) is being developed [6]. In this process shown in Fig. 8.4b, the shear force (F_t) aids three body wear mechanism between the abrasive and the workpiece surface. Further hybridization can be achieved by giving US vibration to the workpiece that increase the material removal rate. In ultrasonic assisted magnetic abrasive finishing (UAMAF) process surface finish is produced by a flexible magnetic abrasive brush in the presence of both magnetic field and ultrasonic vibrations [7].

8.3 Magnetic Field-Assisted Electrodischarge Processes

In EDM process the efficiency of material removal depends on the intensity of electrical discharges i.e. short time current flow through the plasma channel and its stability and physical properties. It is well known that debris accumulation in the machining zone affects the performance and efficiency of EDM process. It was proven, based on theoretical hypothesis and experiments, that these process limitations can be overcome by adding the magnetic field which can stabilize electrical discharges in the plasma channel. This hybrid EDM process is termed *magnetic field assisted EDM* (MFAEDM) [1, 2, 8, 9].

A typical set-up of magnetic field-assisted EDM installed in a standard EDM machine, shown in Fig. 8.5a, includes a special magnetic force-assisted device. It consists of a rotational disc fastened with two symmetric magnets driven beneath the EDM discharge zone by electric motor. As a result, the generated magnetic field exerts a strong normal magnetic force on the motion of working electrode. In consequence, debris particles subjected to both magnetic and centrifugal forces are

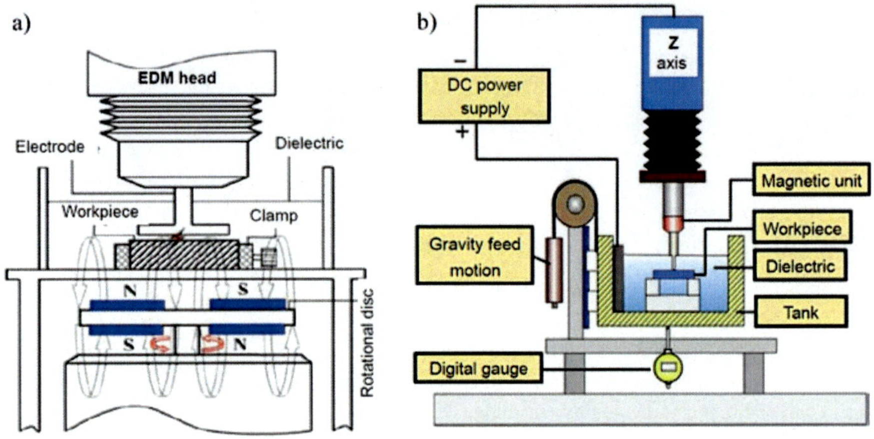

Fig. 8.5 Scheme of magnetic field-assisted electro-discharge hybrid process (MFAEDM) (**a**) and a scheme of the magnetic field-assisted ECDM system (**b**) [8]

effectively flushed from the inter-electrode gap and machining zone. A scheme of the magnetic field-assisted ECDM (MFAECDM) system with a gravity feed drive is shown in Fig. 8.5b. In this hybrid process the magneto-hydrodynamic convection induced by the magnetic field enhances electrolyte circulation in the working zone, which improves machining performance. For instance, the geometrical accuracy achieved in the micro-MFECDM of glass is about 25% higher and machining time is reduced by 60% [8].

It was documented based on measurements of electron density and plasma temperature using spectroscopy methods that the magnetic field stabilizes the behaviour of the plasma channel and reduces the discharge zone (confining). These effects result in higher current density on the workpiece surface and higher efficiency of EDM process, especially in micro-EDM operations [10, 11] (Fig. 8.6).

In EDM sinking of micro-holes an important technological problem is removing droplets of remelted material because they disturb the homogeneous distribution of electrical discharges on the workpiece surface and their performance. When their concentration on the working surface increases the local short-circuits can appear and surfaces of the working electrode and the workpiece can be damaged. As shown in Fig. 8.7, an external homogenous magnetic field localized on the top of the working zone stabilizes the electrical discharges (reduces the probability of abnormal discharges) and causes that they are distributed on the working electrode more homogenously. As a result, the accuracy and efficiency of a hybrid EDM process increase and the surfaces of better quality are produced. It is reported that the material removal rate in the machining of a hard tool steel can be improved even by 300% in comparison to conventional EDM [5]. In addition, the wear of working electrode is reduced [11, 12].

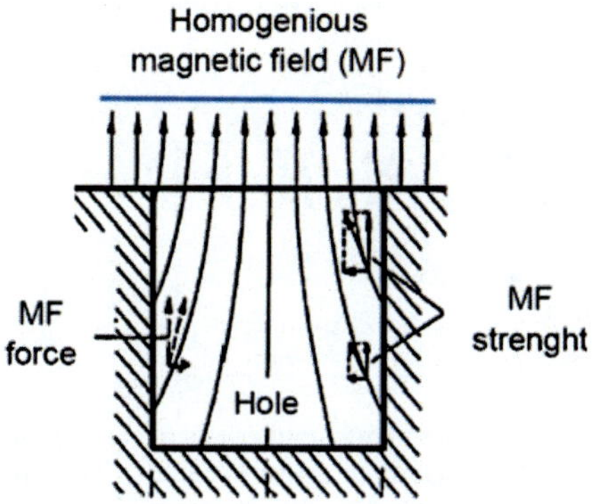

Fig. 8.6 Distribution of magnetic field (MF) lines inside a hole made by EDM sinking and distribution of MF strength and forces acting on the material particles removed from the working zone [8, 12]

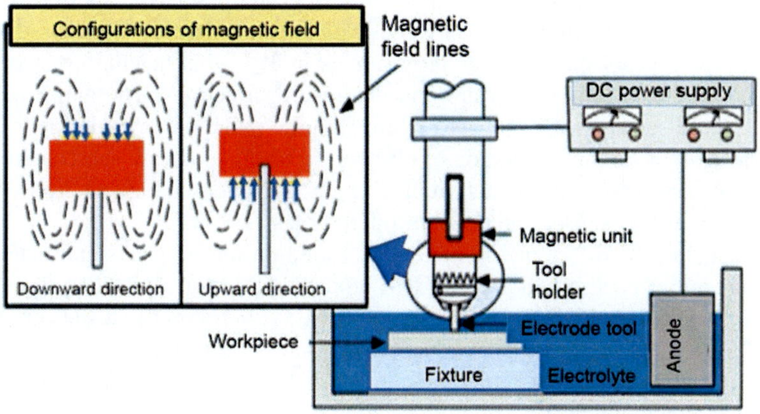

Fig. 8.7 Scheme of magnetic field-assisted ECDM hybrid process (MFAECDM) [13, 14]

8.4 Magnetic Field-Assisted Electrochemical Processes

Pure electrochemical machining process is performed based on the three principle phenomena, i.e. ionization of the workpiece surface, deionization of the working electrode (cathode) surface and ion migration within electrical field. It is evident, that the additional external magnetic field will change all these physical influences and process conditions. It was documented [13] that, for instance, external magnetic field changes the trajectories and velocities of ion migrations, intensifies convection and ion diffusion to the surface and accelerates electrochemical reactions. In consequence, the anodic current density increases and finally smoother surface is generated. This surficial effect is more pronounced for rougher surfaces and the smoothing effect depends substantially on the relationships between characteristics

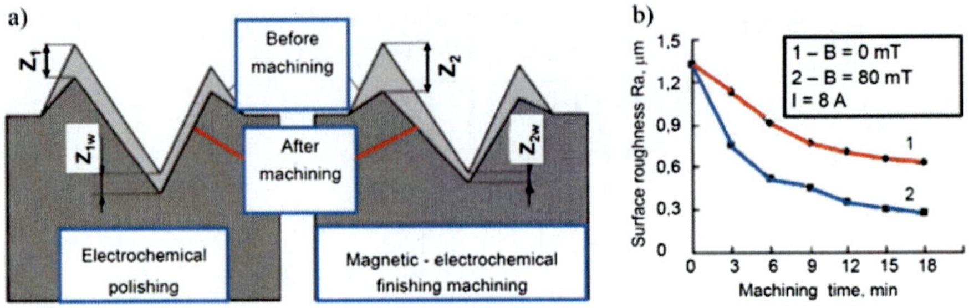

Fig. 8.8 Model for smoothing surface micro-irregularities (**a**) and influence of the MF density on the surface roughness Ra [13]; B—magnetic induction, equals "0" for pure ECM and B = 80 mT (mTesla) for MFAECM process [13]

of both electric and magnetic fields. In pure ECM process ions move along linear trajectories which are changed with the magnetic field assistance (Fig. 8.7) and this fact positively influences the dissolution conditions on the surface asperities [13].

In the device shown in Fig. 8.7 a magnetic unit is placed into a special tool chuck which holds the tool electrode. A provision is built in this machining system to apply the magnetic field in upward and downward directions. The application of magnetic field causes that a magneto-hydrodynamic (MHD) convection is induced in the electrolyte, so the circulation of electrolyte within a narrow gap is enhanced. This phenomenon prevents the debilitation of gas film over the tool electrode which results in better geometrical accuracy of micro holes.

In particular, the difference in material dissolution on the peaks and valleys of surface asperities is observed in pure ECM process but this effect is more significant with the magnetic field assistance (Fig. 8.8a). A visible effect that the roughness height ($Z_{2w} < Z_{1w}$) is confirmed by surface roughness measurements. The comparison of surface roughness parameter Ra in Fig. 8.8b indicates that it progressively decreases down to about 0.3 µm during the process time of about 18 min when the magnetic field with the induction of 80 mT was induced.

Better surface quality can be produced with additional assistance of US vibrations with optimized amplitude and power [15]. The combined hybrid ECM process with the simultaneous assistance of magnetic field and US vibration induced to the workpiece can be effectively performed on the setup shown in Fig. 8.9. This multi-purpose hybrid device is composed of a magnetic mechanism, DC power supply, pulse generators and electrolyte supplying system. It should be noted that it is mostly suited to fabricate 3D micro-shapes on non-conductive materials such as glass and quartz [8].

The machining system shown in Fig. 8.9 can be used for investigations of both MFAECM and USAECM with workpiece vibrating processes. It is capable of producing micro-parts in the following ECM finishing operations:

- With supplying DC and impulse current,
- With the assistance of US vibration,
- With the assistance of magnetic field,
- With the simultaneous assistance of magnetic field and US vibration.

This flexible device has a huge potential in precision fabrication of micro-parts made from special materials (for instance HAP 72 alloy) with a very good finish quality (the achievable surface roughness is Ra << 0.1 μm [15]). This results not only from a wider combination of constituent micro-machining processes but also more accurate selection and optimization of process parameters. It was documented based on experimental investigations that the best technological results are achieved by simultaneous assistance of magnetic field and US vibrations, similarly as in the UAMAF process [8, 15]. For instance the optimal machining conditions used in the hybrid machining of HAP 72 are: $NaNO_3$ electrolyte with 20% concentration and flow rate of 20 l/min; electrolyte temperature of 50 °C, current density of 60A/cm^2, magnetic induction of 3000 G (1 Gauss [G] = 0.0001 T [T), distance between magnet poles of 200 mm [15].

8.5 Electric Field-Assisted Processes

The assistance of an external electric field to micro machining process is an effective method for removing debris and preventing particles from redepositing onto the machined surface. A scheme of the setup for external electric field- assisted femto-second (fs) laser micro-machining is presented in Fig. 8.10a. The ultrashort (fs) laser pulses are obviously used as the most effective method for formation of the plasma cloud but its presence frequently leads to the redeposition of large number of micro/nano-sized particles on the substrate surface. When an external electric field (voltage equal to 150 V) is applied in the fs laser micro-machining of silicon part, as shown in Fig. 8.10b, the redeposition of material particles is sufficiently inhibited.

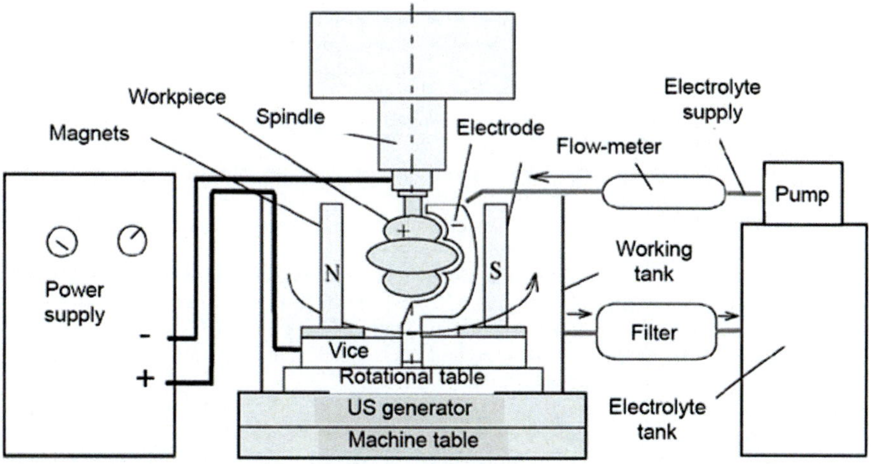

Fig. 8.9 A scheme of the combine vibration and magnetic field-assisted electrochemical machining system [8, 15]

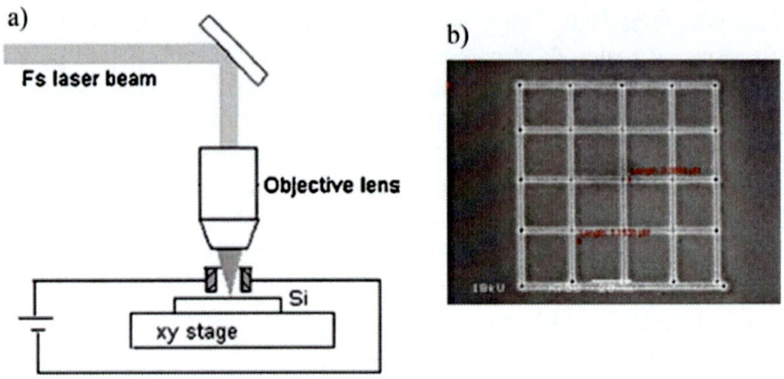

Fig. 8.10 A scheme of the external electric field- assisted femto-second (fs) laser micro-machining (**a**) and example of micro-circuit made in silicon plate [2, 8] (www.home.iitk.ac.in)

Surface is much cleaner in comparison to laser machining without external electric field and part quality is significantly improved (surface contaminations are minimized because only the particles with weight smaller than the electric force can be removed). It was proven that higher voltage produces higher electric forces and attract larger charged particles and facilitates efficient debris removal [2].

References

1. Gupta K, Jain NK, Laubscher RF (2016) Hybrid machining processes. Perspectives on machining and finishing. Springer, Heidelberg
2. Luo X, Qin Y (2018) Hybrid machining, Theory, methods, and case studies. Elsevier, Academic Press, London
3. Jain VK (2013) Manufacturing processes. CRC Press, Boca Raton
4. Jain VK (2009) Magnetic field assisted abrasive based micro-/nano-finishing. J Mater Proc Technol 209(20):6022–6038
5. Heng L, Kim YJ, Mun SD (2017) Review of superfinishing by the magnetic abrasive finishing process. High Speed Mach 3:42–55
6. Kumari Ch, Chak SK (2018) A review on magnetically assisted abrasive finishing and their critical process parameters. Manuf Rev 5(13):1–14
7. Das M, Jain VK (2008) Analysis of magnetorheological abrasive flow finishing (MRAFF) process. Int J Adv Manuf Technol 38:613–621
8. Chavoshi SZ, Luo X (2015) Hybrid micro-machining processes: a review. Prec Eng 41:1–23
9. Lauwers B, Klocke F, Klink A, Tekkaya AE, Neugebauer R, Don Mcintosh D (2014) Hybrid processes in manufacturing. CIRP Ann Manuf Technol 63(2): 561–583
10. Bhowmik S, Divya Zindani D (2019) Hybrid micro-machining processes. Springer Briefs in Applied Sciences and Technology, Berlin
11. Albinski K, Musiol K, Miernikiewicz A, Labuz S, Malota M (1996) The temperature of a plasma used in electrical discharge machining. Plasma Sour Sci Technol 5(4):736
12. Yeo SH, Murali M, Cheah HT (2004) Magnetic field assisted micro electrodischarge machining. J Micromech Microeng 14:1526–1532
13. Fang JC, Jin ZJ, Xu WJ, Shi YY (2002) Magnetic electrochemical finishing machining. J Mater Proc Technol 129:283–287

14. Singh T, Dvivedi A (2016) Developments in electrochemical discharge machining: a review on electrochemical discharge machining, process variants and their hybrid methods. Int J Mach Tools Manuf 105:1–13
15. Pa PS (2009) Superfinishing with ultrasonic and magnetic assistance in electrochemical micromachining. Electroch Acta 54:6022–6027

Chapter 9
Thermally-Assisted Machining Processes

9.1 Classification of Thermally-Assisted Machining Processes

Thermally assisted machining (TAM) is an effective method of cost-effective production of high quality parts by increasing process efficiency, improving dimensional control, reducing componential cutting forces, increasing tool life and reducing surface defects when machining materials such as superalloys and ceramics and when using micro tools with limited strength [1]. So far, acetylene flame, electric current, induction heating (eddy currents), laser and plasma have been used as sources of thermal energy supplied to the workpiece in the machining zone. However, laser and plasma are considered to be the most effective sources of thermal assistance for the machining process. Some cases of TAM are illustrated in Fig. 9.1 Their characteristics are summarised in Table 9.1.

Nowadays, due to the possibility of generating and controlling a laser spot of about 800–1000 °C directly in front of the tool, some types of lasers or plasma arc are used in TAM processes [1–3]. Figure 9.2 shows the possible applications of the laser as an energy source in materials manufacturing, taking into account micromachining and cutting/drilling. Due to the power requirements, laser assisted machining (LAM) uses a CO_2 molecular laser with a wavelength of 10.6 μm, and most commonly, for the reasons given earlier, a pulsed solid-state Nd:YAG laser emitting infrared radiation with a wavelength of 1.064 μm [1, 4]. It is generally required that the continuous wave beam used for heating has a Gaussian normal distribution (Sect. 5.1.4). It should be added that the CO_2 laser is best suited for heating ceramics due to its ideal absorption, but has limitations when a mirror is used to transmit the laser beam.

It is generally required that the thermal energy sources used in TAM machining have the following characteristics [2, 5]:

- High energy density required to rapidly heat the material,
- Easy control of the heating area and its location,

W. Grzesik and A. Ruszaj, *Hybrid Manufacturing Processes*, Springer Series in Advanced Manufacturing, https://doi.org/10.1007/978-3-030-77107-2_9

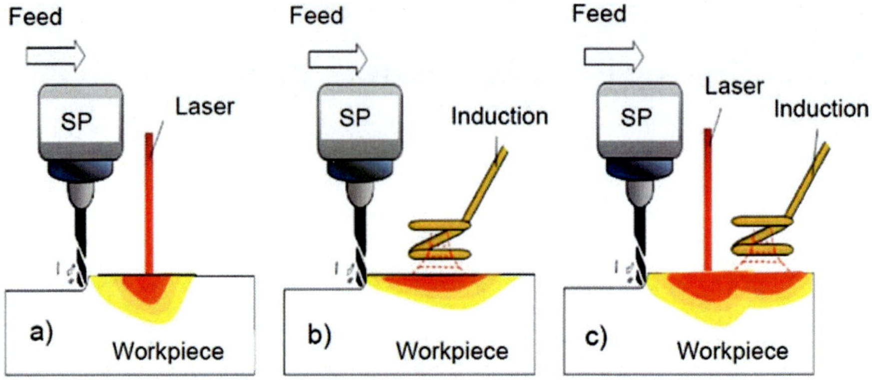

Fig. 9.1 Different heating methods of material layer in front of the tool in the case of milling operation: **a** laser, **b** induction coil, **c** two stepped using induction coil and laser [14]

Table 9.1 Characteristics of sources/methods of heating applied in TAM processes

Heat source	Advantages	Disadvantages
Laser	• High degree of heat concentration • Simple control of heat source	• High equipment cost • Dependence of absorption on material grade
Plasma	• High degree of heat concentration	• Not possible precise control
Rotary current (induction coil)	• Ease service • High heating efficiency	• Not possible high heating concentration • Limited device movement
Electrical current	• Simple equipment • Homogeneous heating	• Not possible precise control
Gas flame	• Low investment cost	• Not possible high heating concentration

– Safety, low cost and the easiest possible adaptation to conventional machines. This is, however, as in milling, made more difficult by the rotating tool.

The previously mentioned sources/methods of material preheating, and the laser in particular, are used in hybrid conventional machining processes—turning, milling, grinding—and unconventional machining processes—EDM, ECM, WJM. Laser-assisted processes will be discussed in detail in Sect. 9.3.

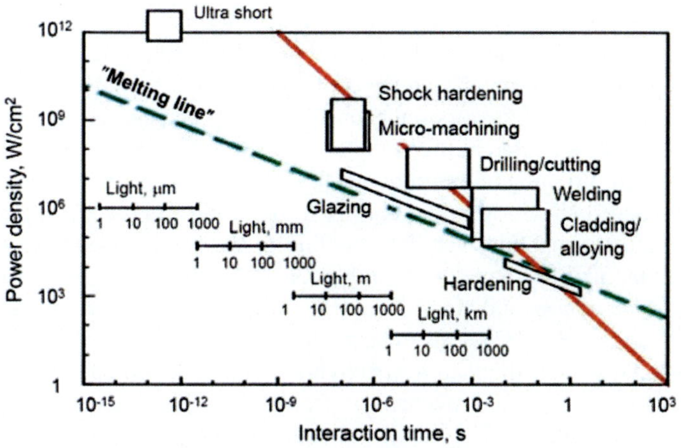

Fig. 9.2 Laser applications in manufacturing according to a typical pulse duration used [1, 4]

9.2 Physical and Technological Effects

In thermally-assisted machining, it is assumed that its effectiveness is assured if the preheating temperature is controlled with respect to the plastic deformation temperature or the occurrence of the metastable phase. For example, each group of nickel-based superalloys has a different deformation temperature depending on the alloying constituents. Strain temperature, strain rate and the effect of heating on the strain rate are fundamental factors in determining which structural change mechanism can control the shear stress value. It appears that choosing the right temperature at which microstructural changes occur can lead to both softening and hardening of the material. In the case of laser assisted machining (Fig. 9.3a), there are cyclical periods of

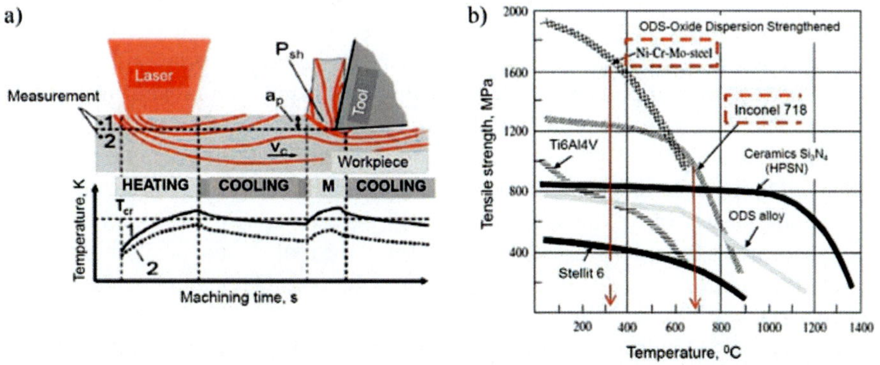

Fig. 9.3 Scheme of thermal influence on material (**a**) and the reduction of material tensile strength (**b**) in laser-assisted machining [1, 2]. Symbols in Fig. 9.3a—H—heating, *COOL*–cooling, *M*—machining, T_{cr}—critical temperature

heating and cooling due to the location of the heat source at a certain distance in front of the tool. It should also be kept in mind that the pulsating, wave-like laser beam can harden the material during the cooling phase and have the opposite effect during the process. This is an important consideration to take this effect into account when selecting a heat source to thermally assist the process. Figure 9.3b shows the results of thermal softening of different metallic (Ti–6Al–4V titanium alloy, Inconel 718 alloy, hardened Ni–Cr–Mo tool steel used for mould and die making and ODS type alloy) and non-metallic materials (Stellite 6 and Si_3N_4 nitride ceramics). In the case of titanium alloy Ti–6Al–4V and Inconel 718 alloy, a visible thermal softening is observed at 600–700 °C. However, for Ni–Cr–Mo tool steel, preheating to temperatures of 250–350 °C seems to be sufficient.

The beneficial effects obtained from a hybrid TAM process depend on the material grade and its thermal and mechanical properties. When metallic materials are machined, e.g. austenitic steels, Inconel 718, titanium alloys, metal-ceramic composites, localised preheating causes a reduction in the yield stress lesser strain hardening effect, resulting in lower energy consumption and improved machinability. In the case of thermally-assisted machining of structural ceramics, an intense heat source (laser, plasma, electric arc, etc.) locally heats the workpiece above the critical temperature in order to reduce the yield strength of the material below the brittle fracture strength, so that material removal takes place in a quasi-plastic state rather than in a brittle state.

9.3 Laser-Assisted Machining (LAM)

9.3.1 Laser-Assisted Cutting Processes

As mentioned in Sect. 9.1, laser assistance is most commonly used in turning, less frequently in milling and grinding operations. It can be used for both conventional machining and micromachining [2, 5]. A typical solution for turning is to mount the LAM system on a gantry above the CNC machine tool, so that it is possible to locate the beam at a selected point on the workpiece surface and then move it in front of the tool. The axial and circumferential positions of the laser beam heating the cut layer are shown in Fig. 9.4a and b respectively. The distance L_1 between the focusing lens and the object results from the assumed laser spot size, so as to cover the entire cross-sectional area of the transient surface. The distance L_2 (Fig. 9.4b), on the other hand, determines the shift of the heating in front of the tool, i.e. the positioning of the spot on the circumference of the workpiece and the possible heating of the cutting tool material.

Good heating effects are obtained when the cutting tool is made of mixed (Al_2O_3 + TiC) ceramics with very low thermal conductivity [1, 2, 5]. The laser-assisted turning scheme used for machining nitride ceramics and partially stabilised zirconia ceramics (PZS) is shown in Fig. 9.4b. It can be seen that, due to the low thermal

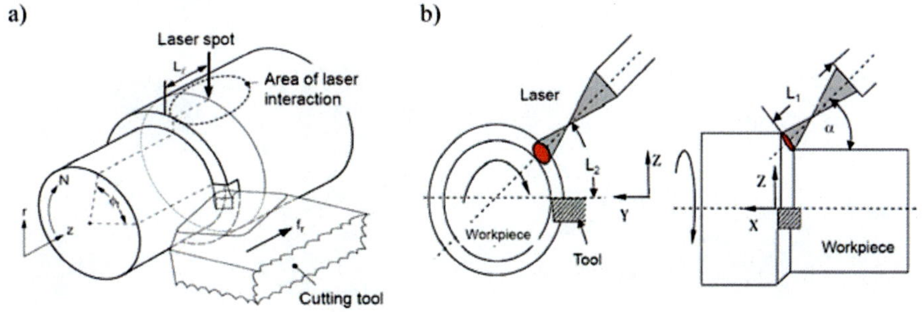

Fig. 9.4 Localization of laser beam in material heating by a single heat source on: **a** material circumference and **b** transient surface [1, 6, 7]

diffusivity of these materials ($\alpha = 8.6 \times 10^{-6}$ m/s^2 for Si_3N_4 ceramics and $\alpha = 1.1 \times 10^{-6}$ m/s^2 for PSZ), the laser focus is shifted axially and circumferentially with respect to the tool edge in order to gain the time necessary to concentrated heating the workpiece material to the required temperature. It should be added that laser assisted turning (LAT) can be competitive to grinding, since the specific energy of cutting (precisely, its mechanical part) is only 6–16% of the total energy. However, the efficiency of laser heating needs to be improved.

In thermally assisted turning, a single source (Fig. 9.4a, b) or two (Fig. 9.6a) laser sources are used. In the first case, the spot is located on the top of cylindrical surface at a distance L_l from the cutting edge or perpendicular to the transient surface keeping additionally an appropriate angular displacement. In the second case (Fig. 9.6a), the both laser sources are used simultaneously. The third possible arrangement (Fig. 9.5b) is direct heating of the cutting zone (*DLAM-direct LAM*) [8], when the laser beam passes through a transparent sapphire tool, with a micro texture applied on it to facilitate chip formation. This method has been successfully validated in the cutting of Zr-BMG (*Bulk Metallic Glass*).

In modelling the effect of laser preheating, not only the material removal temperature T_{mr}, measured on the shear plane (the shear plane model is described in Sect. 3.1.1), but also the temperature distribution in the cut layer at the depth equal to a_p value is important (Fig. 9.3a). The influence of heat dissipation due to friction, which has a significant contribution in the classical thermal model of the cutting process (Sect. 4.3), is omitted. It should be added that laser heating introduces very large temperature gradients in the cross-section of the machined layer, which is illustrated by a thermogram recorded with an infrared camera (Fig. 9.6b). A schematic of the heat transfer model in LAT is shown in Figs. 9.4a and 9.5a. In practice, empirical equations can be used to determine the process temperature as a function of the LAT process conditions for specific materials. Example equations have the following form [2]:

– for Inconel 718 alloy and two CO_2 and Nd:YAG lasers

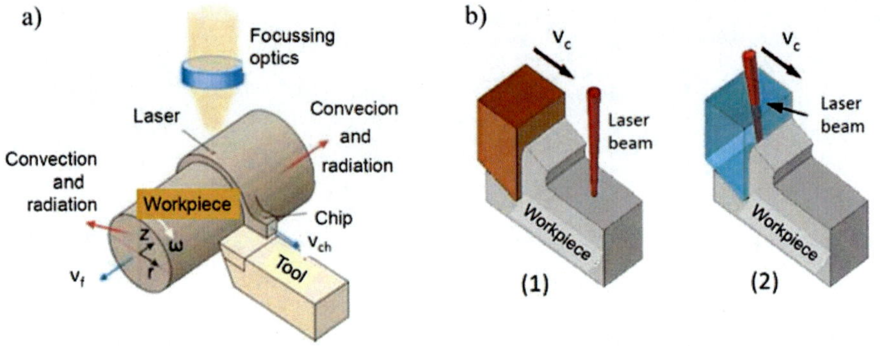

Fig. 9.5 Heat transfer conditions in the model of material heating by one laser beam in turning (**a**) and example of direct laser material heating—direct laser-assisted machining (DLAM) (**b**) [1, 2, 8]

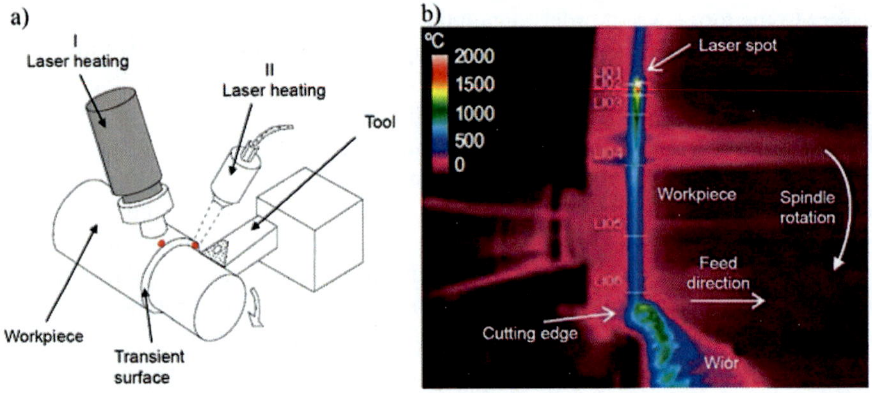

Fig. 9.6 Principle of turning process assisted by two laser sources (**a**) thermograph of the cutting zone (**b**) [6, 7]

$$T_{mr} = 27890\frac{P_{CO_2}^{0.086}P_{YAG}^{0.031}}{D_W^{0.95}f^{0.32}} \tag{9.1a}$$

– for titanium alloy Ti–6Al–4V and one CO_2 laser

$$T_{mr} = \frac{e^{3.4}P_{CO_2}^{0.66}}{D_W^{0.34}f^{0.31}v_c^{0.31}} \tag{9.1b}$$

where: *P*—laser power, D_W—workpiece diameter, *f*, v_c—machining parameters (feed rate and cutting speed).

LAMILL (Laser Assisted Milling) is another hybrid machining process that combines material pre-heating with mechanical material removal. Unlike turning

(Fig. 9.5a), there are problems with the installation of the laser source due to the rotational tool (possible solutions are discussed in Sect. 9.5). In principle, the laser can be located outside the tool (Fig. 9.7a) or integrated with the machine spindle (Fig. 9.23a).

For face milling it is easiest to locate the laser beam in front of the tool moving in the feed direction as shown in Fig. 9.8. However, due to the limited spot size the heating cannot cover the entire width of the cut layer. It is used to position the laser at a central position (1), at the point of cutter engagement (2), or in both positions (1) and (2) simultaneously. Pre-heating along the tool-workpiece contact (4) is possible when using a laser with a sufficient power, e.g. a 14 kW CO_2 laser for machining titanium and Inconel 718 [6].

Due to the intermittent nature of the milling process, an important issue is the prediction of the temperature distribution on the uncut (upper) and cut (lower) surfaces as shown in Fig. 9.9a. The temperature distribution is determined over the

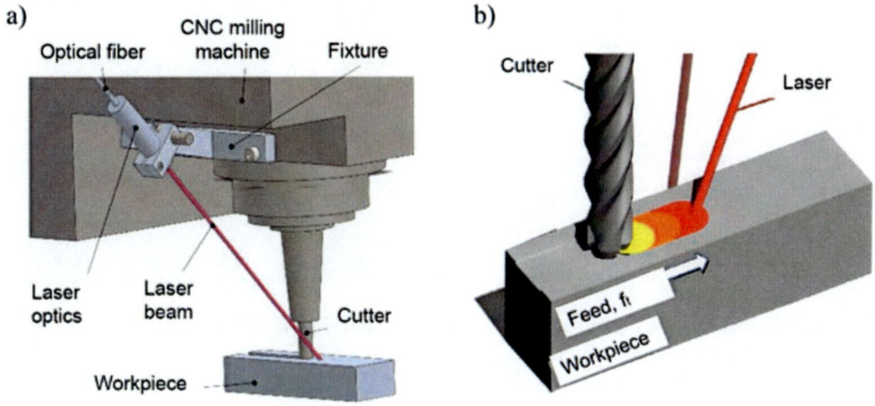

Fig. 9.7 Concept of experimental system of laser-assisted milling process (**a**) and localization of laser spot before the milling cutter (**b**) [9]

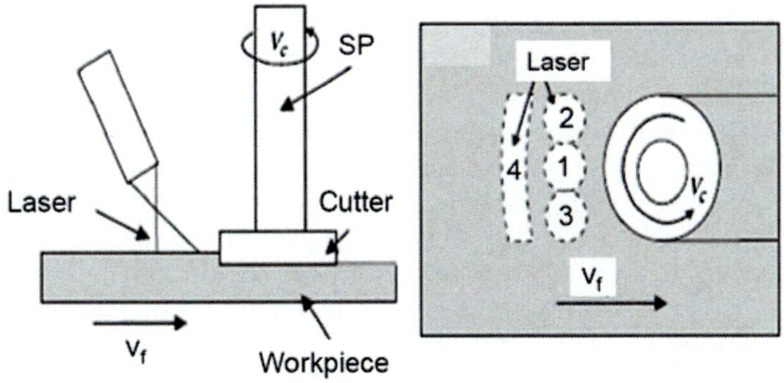

Fig. 9.8 Principle of integration of laser spot and milling cutter positions [5, 7]

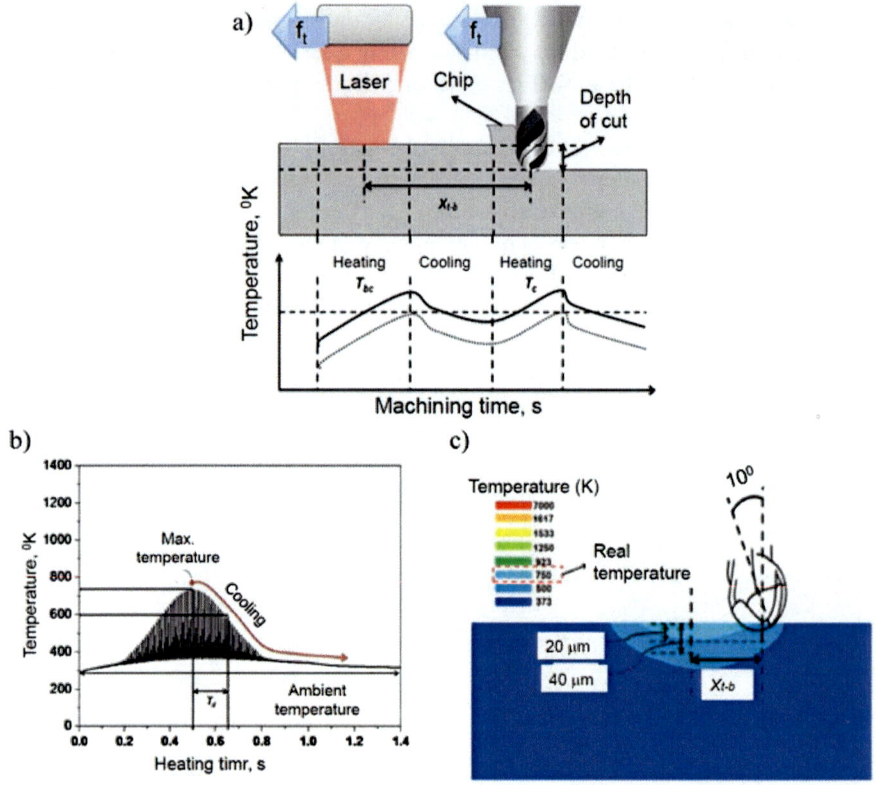

Fig. 9.9 Thermal model of LAMILL process (**a**), distribution of temperature measured at a distance between the cutter engagement and laser spot position (**b**), FEM simulation using ANSYS APDL programme (**c**) [12] (www.intechopen.com)

distance $x_{t\text{-}b,}$ which is covered by the milling cutter from the engagement point to the area heated by the laser. In the simulation, $x_{t\text{-}b} = 600$ μm was assumed to avoid the influence of partial heat flow into the cutting tool with T_c temperature and the initial workpiece temperature of $T_{bc} = 700$ °C. This is consistent with the findings in the LAM process that at 650–900 °C there is a pronounced thermal softening and reduction in material strength (Fig. 6.3b).

Figure 9.10 shows a scheme of laser assisted micro-milling (LAMM) of a slot and the simulated temperature distribution in the machining of titanium alloy Ti–6Al–4V along the L_x section from the cutter axis to the central point of the ellipse-shaped laser spot. It was found that the temperature along the groove rounding T_{mr} changes from 400 do 580 °C when the cutter rotates from 0° to 90°. Laser assisted LAMM machining is used to avoid the formation of burrs on the edges of the groove and to improve the surface roughness.

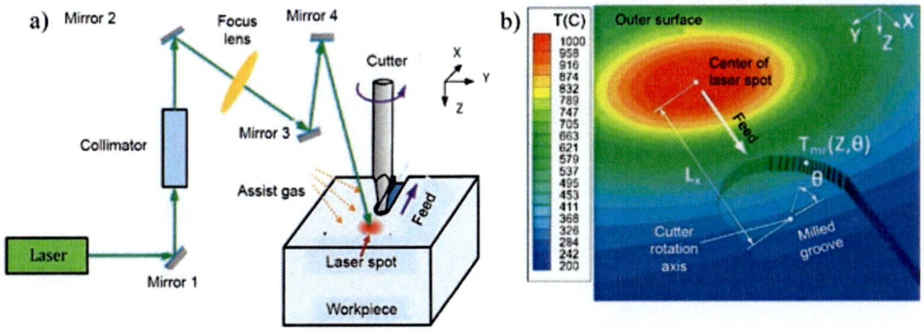

Fig. 9.10 Scheme of optical system in laser assisted micro-slot milling (**a**), temperature distribution at the distance between the cutter axis and laser spot centre (**b**) [10]

9.3.2 Laser-Assisted Grinding Processes

Figure 9.11a shows a scheme of laser-assisted grinding (LAG), which arrangement is essentially the same as in turning and milling discussed earlier. Figure 9.11b, on the other hand, shows a similar scheme for laser-assisted micro-grinding (LAMG). Both processes are mainly used to produce micro-fragments in components made of difficult-to-machine materials such as Al_2O_3 oxide ceramics, Si_3N_4 nitride ceramics and SRB-Sintered Reaction Bonded SiC ceramics (SRB-SiC), where the primary technological problem is to eliminate surface and subsurface micro-cracks that occur after conventional grinding or after prior laser heating.

The laser preheating conditions for these materials are shown in the diagram in Fig. 9.3b. An important problem is to match the laser wavelength with the heat absorption capacity of these ceramic materials. For example, a diode laser with a wavelength of 808 nm provides an absorption of 75% in the heating of Al_2O_3

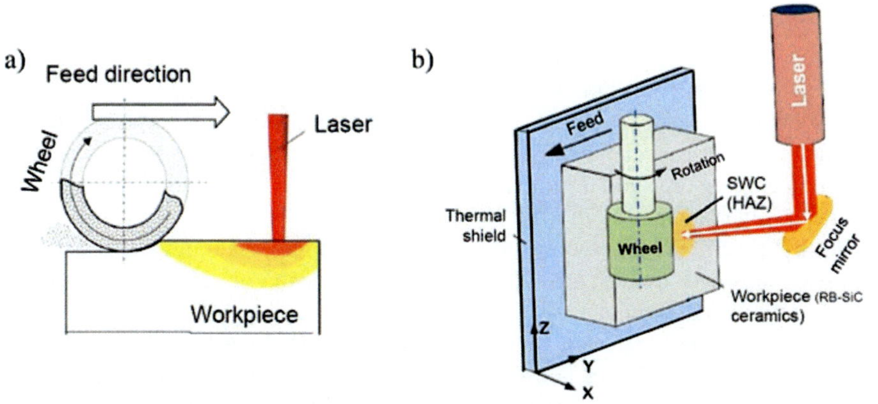

Fig. 9.11 Scheme of laser assisted grinding process (**a**) and micro-grinding (**b**) [13, 14]. HAZ—*Heat affected zone*

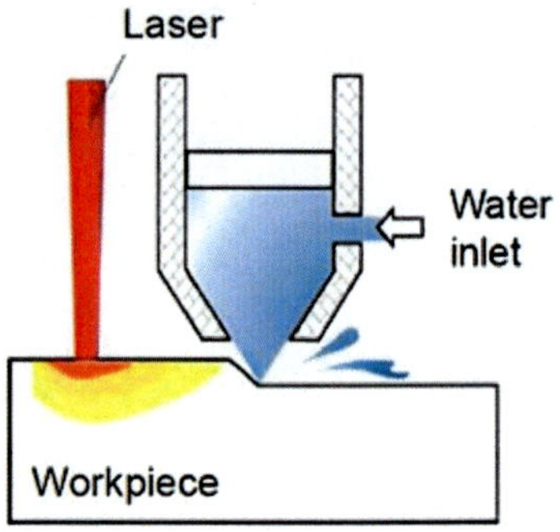

Fig. 9.12 General concept of laser assisted WJM process [13]

ceramics [2]. Micro-grinding of SiC ceramics (Fig. 9.11b) is carried out on an ultra-precise Micro-3D micro-machine tool. The workpiece is fixtured in a clamping device insulated with a heat shield. A fibre laser with a wavelength of 1064 nm is used for heating. A focusing lens is mounted to the spindle and connected to the laser via an optical fibre, allowing free movement between the 3 mm diameter laser spot and the workpiece.

9.3.3 Laser-Assisted WJM Process

Figure 9.12 shows a rather original concept for the laser assisted WJM process. For this purpose, a CO_2 laser is used as a heat source for material pre-heating. Improved surface quality and a clear increase in productivity are achieved [13].

9.3.4 Laser-Assisted EDM Processes

Laser-assisted EDM (LAEDM) is a hybrid process in which laser machining (LBM) and electrodischarge machining (EDM) take part as sequential processes according to the scheme shown in Fig. 9.13a, b. First, the initial outline is made by laser ablation and then μ-EDM is used for finishing. The main limitations of this method are the longer machining time and the intensive electrode wear in EDM and the appearance of a recast layer and a heat affected zone in LBM.

The hybrid LAEDM method is mainly useful for micromachining 3D structures, grooves, holes and cavities, e.g. 140 μm diameter holes in diesel injection nozzles to reduce time and eliminate the negative effects of the laser, i.e. defects in the re-melted layer and heat affected zone (HAZ) through EDM machining (Fig. 9.13c). Typically, an Nd-YAG laser is used. The process parameters selected include the EDM voltage, electrode type and material, dielectric grade, laser impulse power and impulse frequency.

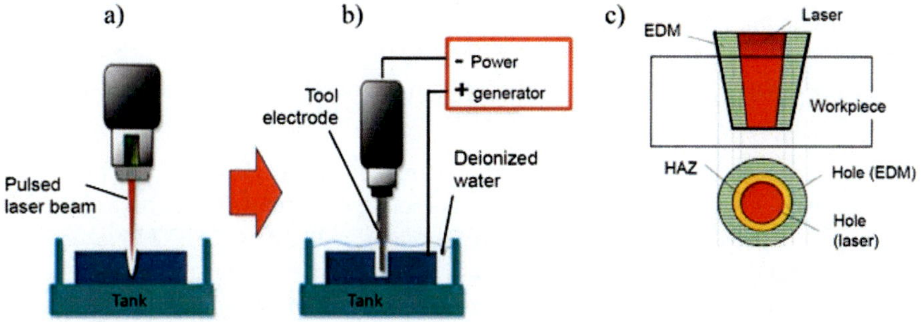

Fig. 9.13 Scheme of laser-assisted micro-EDM (μEDM) process: **a** initial operation using laser ablation, **b** finish operation using μEDM, **c** removing HAZ in EDM [15]

9.3.5 Laser-Assisted ECM Processes

When shaping components with demanded good surface quality, it is reasonable to use electrochemical machining. For many materials, however, this process is difficult to carry out due to the increase in passivation and the difficulties associated with the localisation of the dissolution process. A rational solution to this problem is to heat the machining zone with a laser beam and perform *laser assisted ECM* (LAECM). In the LAECM process, the main mechanism of material removal is electrochemical dissolution [16] while the laser beam increases the temperature in a specific machining zone and thus reduces the intensity or eliminates the passivation effect (Fig. 9.14).

The main effects of the laser beam are [16]:

- Increase in the rate of electrochemical dissolution by increasing the temperature at inter-electrode gap according to Arrhenius' law,

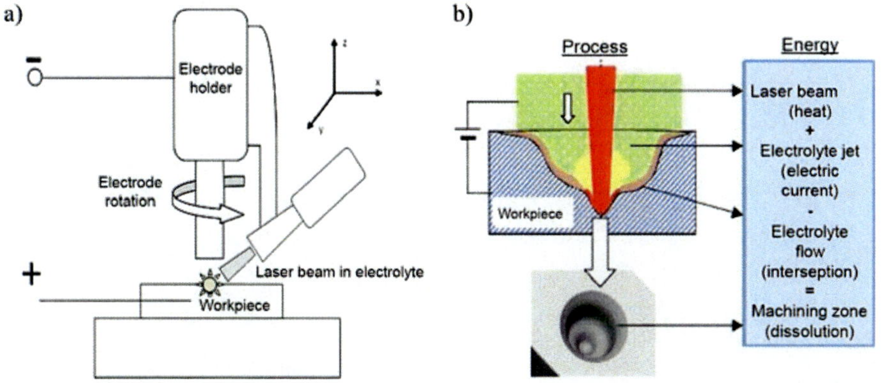

Fig. 9.14 Concept of machine for LAECM process (**a**) scheme of energy supply to the machining zone (**b**) [3, 16] (www.intechopen.com)

- Increase in the temperature of the electrolyte and thus its conductivity and current density,
- Elimination of surface passivation, which favourably increases the material removal rate.

The concept of the device for laser assisting of electrochemical machining is shown in Fig. 9.14a. The main components included in the workstation are the X, Y worktable, Z-axis drive, ECM impulse generator and 532 nm Nd:YAG laser. The above test stand concept was developed based on mathematical modelling of the LAECM process. The modelling results indicate that increasing the temperature in the machining zone is sufficient to intensify the ECM process (depassivation, depolarisation, improved localisation of the ECM process). Modelling results also indicate that one of the basic parameters of the LAECM process is the beam incidence angle, which should be in the range of 0°–20° (Fig. 9.14b). In LAECM low power lasers, e.g. 375 mW, are used. As already mentioned, the laser beam raises the temperature in the machining zone, which improves the localisation and efficiency of the process and initiates the phenomena of depassivation and depolarisation [17, 18].

A scheme of the LAJECM (laser-assisted jet electrochemical machining) working zone is shown in Fig. 9.15a. The three characteristic areas A, B and C (Fig. 9.15b) are related to the following interactions:

- Zone A is the concentrated area of the laser operation. The marked vectors roughly represent the material dissolution rates in the laser beam cross section,
- Zone B is the area where the material is dissolved by a portion of the electrolyte jet. The vectors also roughly represent the material dissolution rates,
- Zone C is the area of interaction between the laser and the electrolyte and therefore a higher concentration of dissolved material particles.

A schematic diagram of a test stand for LAJECM machining is shown in Fig. 9.16. Examples of test results for the dependence of material removal rate on the inter-electrode voltage and inter-electrode gap width (test conditions: electrolyte concentration of 10–20%, electrolyte flow velocity of 15–19 m/s, average laser power of

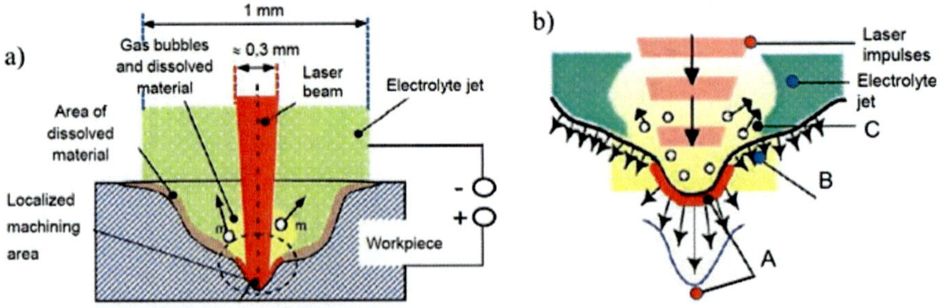

Fig. 9.15 Scheme of machining zone **a** and characterization of physical phenomena in the localized working zone **b** for laser-assisted jet electrochemical machining (LAJECM) [3, 17, 18] (www.home.iitk.ac.in)

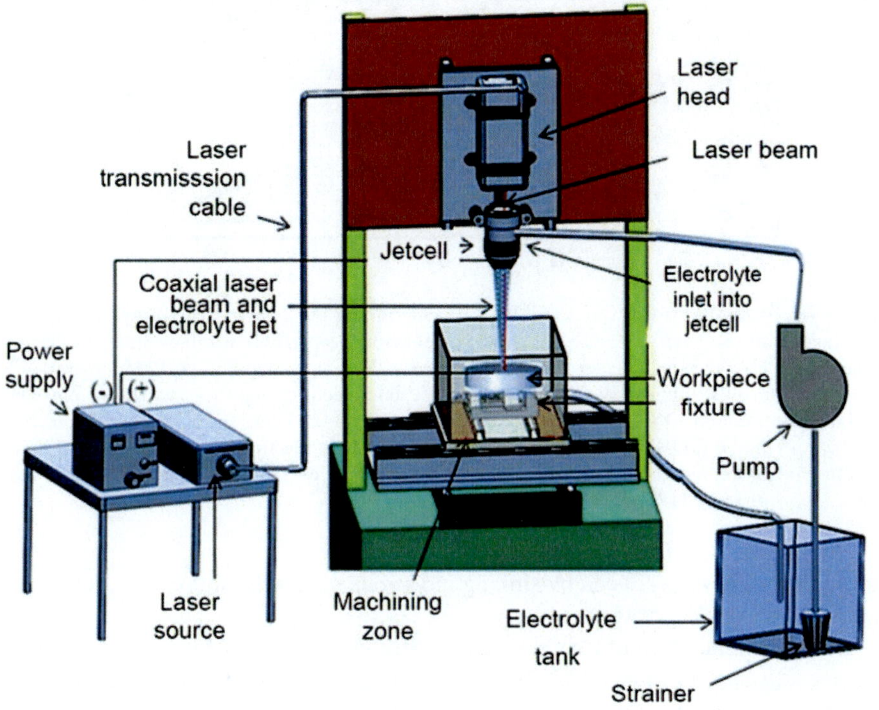

Fig. 9.16 Scheme of the LAJECM process test stand [17, 18]

375 mW, impulse frequency of 10–15 Hz, laser spot diameter of 0.1–0.3 mm, laser radiation power density equal to 47.5 W/mm^2) are presented in Fig. 9.17. In turn,

Fig. 9.17 Dependence of machining efficiency on inter-electrode voltage and gap width (IEG-Inter—Electrode Gap) [17, 18]

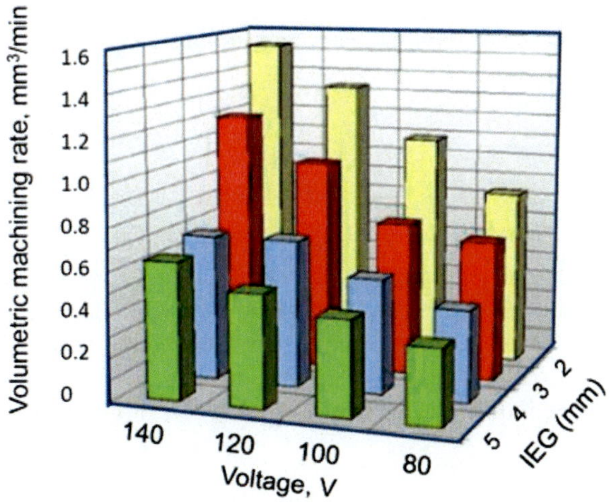

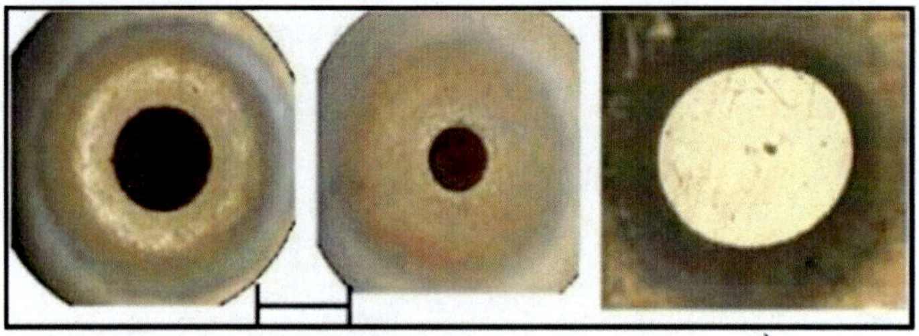

Fig. 9.18 Examples of hole produced by different sinking processes—magnification 50×: **a** in aluminium alloy—LAJECM: U = 120 V, interelectrode gap (IEG) width—4 mm, electrolyte concentration 10%, hole diameter for beam entrance—0.31 mm, and at exit—0.12 mm; hole depth: 1.45 mm; **b** in aluminium alloy—JECM: U = 120 V, interelectrode gap (IEG) width—4 mm, electrolyte concentration 10%, hole diameter for beam entrance—0.29 mm, and at exit—0.08 mm; hole depth: 1.45 mm; **c** LAJECM in stainless steel: U = 180 V, interelectrode gap (IEG) width—2 mm, electrolyte concentration 10%, hole diameter for beam entrance—0.31 mm, and at exit—0.20 mm; hole depth—0.5 mm [17, 18]

examples of holes produced by sinking in various materials and process conditions are presented in Fig. 9.18.

The beneficial effect of laser assistance on the electrochemical machining of materials results, among others, from the change of equilibrium potential, increase in the rate of electrochemical reactions, increase in the rate of mass transfer in the cathodic diffusion layer, decrease in anode passivation or increase in the number of current carriers in the case of semiconductor machining. In electrochemical machining of n-type semiconductors (e.g. β- SiC), the material removal rate is limited by the small number of current carriers. Under the influence of laser radiation, photogeneration of these carriers takes place and the magnitude of the saturation current increases many times (even up to several tens of amperes), which enables a significant increase in the anodic dissolution rate. For this reason, laser-assisted electrochemical machining offers wide technological opportunities, hence the implementation of development research [17–20].

9.4 Plasma-Assisted Machining (PAM) Processes

A plasma jet can be used to preheat and improve the machinability of materials such as glasses, ceramics, sintered high speed steels, stainless steels and alloy steels, thus carrying out plasma assisted machining (PAM) [1]. When using plasma as a heat source to preheat the workpiece, the problem of controlling the heating stability in the required zone arises (Table 9.1). In PAM (also *PEM-Plasma Enhancement*

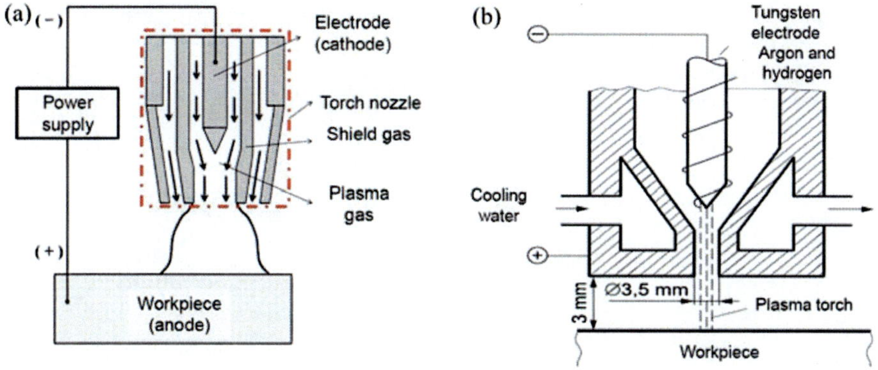

Fig. 9.19 Principle of plasma generation (**a**) and scheme of plasma torch (**b**) [1, 12]

Machining) [12]) schematically illustrated in Fig. 9.19a, a plasma arc is produced by the reaction between the electrical discharges at the interface between the tungsten electrode (cathode) and the gas. If the material is conductive then the nozzle has a negative polarity, i.e. it is a cathode. The opposite is true when the material is non-conductive, e.g. ceramic.

Figure 9.19b shows the working principle of a plasma torch in which the plasma is generated between a tungsten electrode and a copper nozzle, cooled by water into which a mixture of argon and hydrogen is supplied. The gas is ionised by high frequency discharges and the addition of hydrogen allows the temperature on the surface of the workpiece to rise rapidly to about 1200 °C, whereas the temperature at the plasma arc spot reaches values of 16,000–30,000 °C.

In the case of heating hard-to-machine materials, a high energy concentration with a low shield gas flow rate produces a plasma with a maximum temperature in the spot of the order of 16,000 K [1, 12]. However, it should be noted that LAM process is inefficient in terms of energy consumption because it is much higher than in conventional machining. In PAM, which is usually used as plasma-assisted turning, about 30% reduction in the cutting force and about 40% increase in tool life is achieved. In end-milling operations, application a relatively low a workpiece feed rate can result in local melting by the plasma arc.

It has been shown [1] that the use of PAM for machining SS410 stainless steel results in a 20–25% reduction in power and an 80% increase in the MRR. In the machining of Inconel 718, a significant reduction in hardness is obtained already above the temperature of 650 °C, as confirmed by the marked decrease in the resultant cutting force shown in Fig. 9.20. In fact, the force decreases by about 30% when the heated surface temperature is about 500 °C. The absence of burrs and the improvement in surface quality is also worth noting.

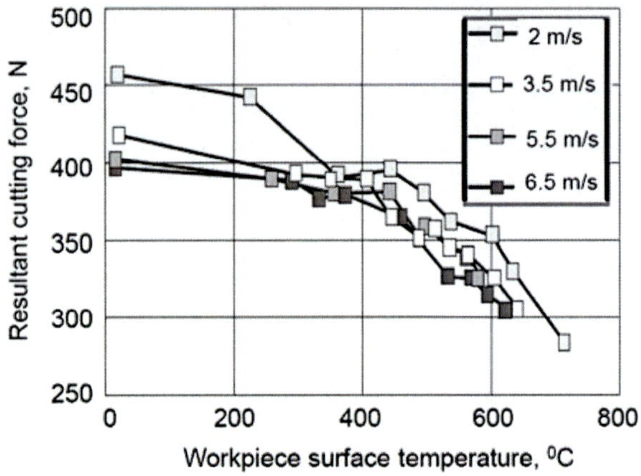

Fig. 9.20 Influence of temperature during plasma heating on the resultant cutting force in machining Inconel 718 alloy [1]

9.5 Industrial Applications of Thermally-Assisted Machining Processes

Thermally assisted machining (TAM) already finds multiple applications in industry, both in conventional machining (mainly laser assistance in turning and milling (LAM) as in Fig. 9.21) and micro-machining (mainly laser assistance in milling and grinding (LAMM)). Turning is the simplest machining operation from a technical point of view, while in milling the problem of integrating the laser with the rotating tool remains open.

The main challenge for this group of hybrid processes is to achieve adequate thermal softening of the material due to preheating and thus its influence on the working temperature of the tool. In experimental studies it was determined that the

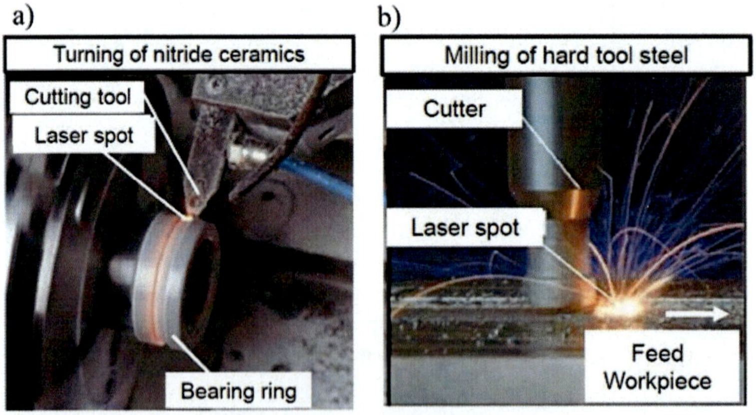

Fig. 9.21 Application of LAM process in turning of bearing ring race made from ceramics (**a**) in milling of a heat resistant steel (**b**) [1, 3]

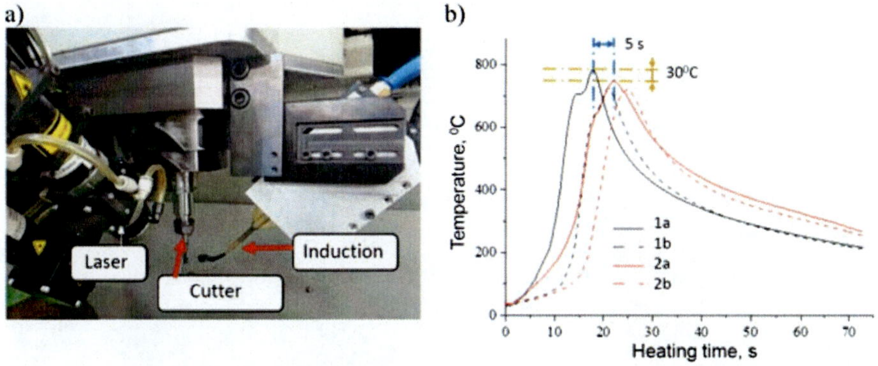

Fig. 9.22 Application of double heat source for preheating (**a**) diagram of temperature distribution for one heating cycle in milling Inconel 718 alloy and feed rate of ft = 200 m/min (**b**) [21]. Symbols: 1—induction plus laser, 2—laser plus induction, **a** cutting depth of 0.1 mm, **b** cutting depth of 0.5 mm

preheating temperature for Inconel 718 (approximately 800 °C) should be higher than for titanium alloy Ti–6Al–4Vs (approximately 600 °C), and therefore a combination of laser (LAM) and induction (*IAM-Induction Assisted Machining*) heating, according to the principle shown in Fig. 9.1c, can be an effective solution. This is important in machining with higher feed rates, when the efficiency of heating with a single heat source decreases dramatically [21]. A design solution for two-stage preheating and the resulting temperature distribution in one heating cycle is shown in Fig. 9.22.

Examples of other designs for laser devices mounted on milling machines are shown in Fig. 9.23. The solution in Fig. 9.23a uses a conventional HSK-100C holder on a 5-axis machining centre, which has an integrated ytterbium fibre laser (λ = 1070 nm) of 2.4 kW focused by two mirrors. The solution in Fig. 9.23b optionally uses two laser beam focusing principles. The first solution (1) uses a fixed laser position and a rotating machine table with precise positioning of the tool path. In the second solution (2), the laser source is rotated around the tool axis (spindle) and positioned in front of the tool. In the solution adjusted to thermal support of the turn-milling process (Fig. 9.23c) a rotary-tilting table controlled in two axes is used.

The solution with a rotating laser source (Fig. 9.24a) is more effective than using a rotary table because it does not influence the feed rate or restrict the cutter's operation. In addition, the focused laser spot can be programmed to oscillate across the width of the removed material, as in Fig. 9.24b. This ensures that the heating spot is concentrated closer to the tool and that a rotating laser source is used when larger diameter tools are used. The disadvantages are the possibility of damaging the laser optics or fibre when rotating them and the high cost of the device [11].

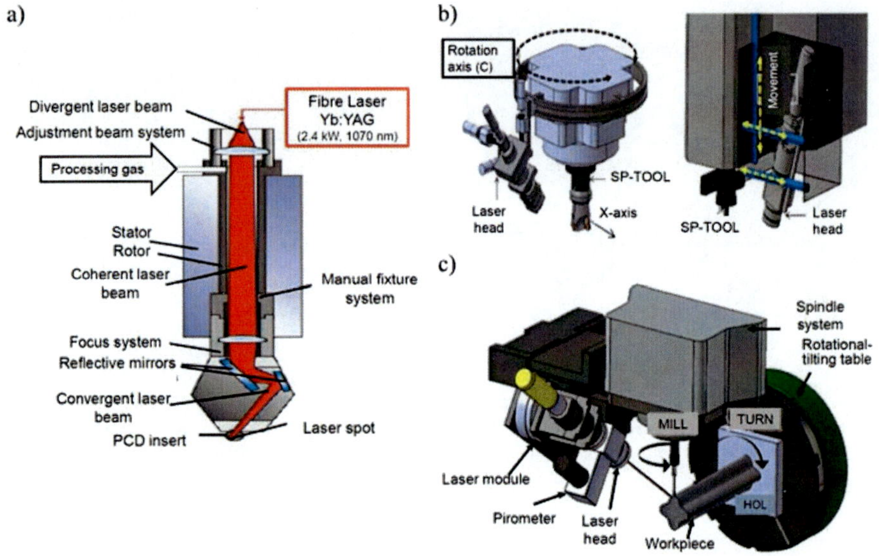

Fig. 9.23 Adaptation of milling centres to LAM: **a** spindle with integrated laser, **b** special bifunctional laser module, **c** device for turn-milling [10, 22]. Symbols: MILL—milling, TUR—turning, HOL—workpiece holder system

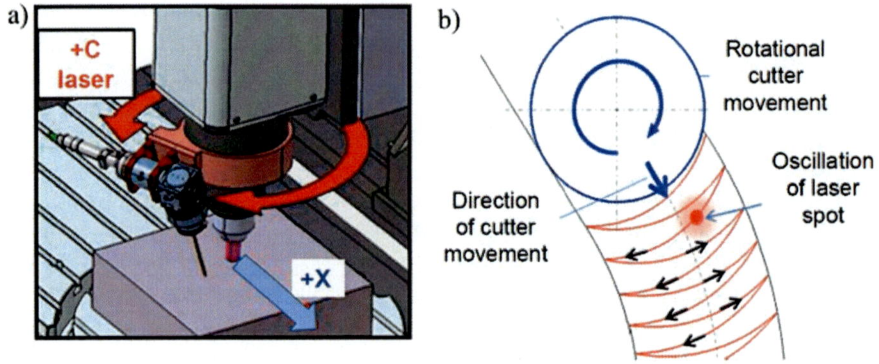

Fig. 9.24 View of controlled axes for application of rotational laser head from Fig. 9.23b (**a**) and programming of trajectory of laser head (**b**) [11]. $+C$—optical axis, $+X$—mechanical axis

It turns out that thermal assistance of cutting and grinding processes is necessary to carry out effective machining of a large group of metallic materials—hardened tool steels, austenitic stainless steels, heat-resistant alloys, and non-metallic materials—ceramics, glasses and composites. An important issue for this group of hybrid machining processes is to ensure an increase in productivity and surface quality.

Among others, TAM processes proved to be effective machining techniques for titanium and nickel-based aerospace alloys with respect to machinability criteria such as tool life, surface roughness and mechanical loads (componential cutting forces) [2, 23, 24] .

References

1. Grzesik W (2018) Advanced machining processes of metallic materials. Elsevier, Amsterdam
2. Luo X, Qin Y (2018) Hybrid machining. Academic Press, (Elsevier) London
3. Lauwers B, Klocke F, Klink A, Tekkaya AE, Neugebauer R, McIntosh D (2014) Hybrid processes in manufacturing. CIRP Ann Manuf Technol 63(2):561–583
4. Meijer K, Du K, Gillner A, Hoffman (2000) Laser machining by short and ultrashot pulses. State of the art and new opportunities in the age of the photons. CIRP Ann 49(2): 523–545
5. Chavoshi S, Luo X (2015) Hybrid micro-machining processes: a review. Prec Eng 41:1–23
6. Ramkumar J, Singh G (2018) Laser-assisted machining. Micromanufacturing Lab., IIT Kanpur
7. Sun S, Harris J, Brandt M (2008) Parametric investigation of laser-assisted machining of commercially pure titanium. Adv Eng Mater 10:565–572
8. Park S, Wei Y, Jin X (2018) Direct laser assisted machining with a sapphire tool for bulk metallic glass. CIRP Ann Manuf Technol 67(1):193–196
9. D. Kramar (2018) Hybrid manufacturing. In: Proceedings of the 3rd international conference on the Industry 4.0. Model for advanced manufacturing. Springer, Berlin, pp 1–17
10. https://research.engineering.uiowa.edu/ding/Micro-machining (2020) LAMM slotting of difficult-to-machine alloys. Laser Materials Processing Lab
11. Lee C-M, Kim D-H, Baek J-T, Kim E-J (2016) Laser assisted milling device: A review. Int J Prec Eng Manuf- Green Technol 3(2):199–208
12. Bermingham M, Schaffarzyk P, Palanisamy S, Dargusch M (2014) Laser-assisted milling strategies with different cutting tool paths. Int J Adv Manuf Technol 74:1487–1494
13. Rahim E, Warap N, Mohid Z (2015) Thermal-assisted machining of nickel-based alloy, open access paper. https://doi.org/10.5772/61101 www.intechopen.com
14. Lee Ch-M, Woo W-S, Kim D-H, Won-Jung Oh W-J, Nam-Seok Oh N-S (2016) Laser-assisted hybrid processes: a review. Int J Prec Eng Manuf- Green Technol 17(2): 257–267
15. Lia Z, Zhang F, Luo X, ChangW CY, ZhongW DF (2019) Material removal mechanism of laser-assisted grinding of RB-SiC ceramics and process optimization. J Eur Ceram Soc 39:705–717
16. Gupta K, Jain N, Laubscher R (2016) Hybrid machining processes. Perspectives and finishing. Springer, Heidelberg
17. Grabowski M, Wyszyński D, Ruszaj A, Skoczypiec S (2011) Application of DPSS laser beam to improve condition of electrochemical processes. In: Proceedings of the INSECT 2011 Conference, pp 91–97
18. De Silwa AK, Pająk PT, Harrison JA, Mc Geough JA (2004) Modelling and experimental investigations of laser assisted jet electrochemical machining. CIRP Ann 53(1):179–182
19. De Silwa AK, Pająk PT, Mc Geough JA, Harrison JA (2011) Thermal effects in laser assisted jet electrochemical machining. CIRP Ann Manuf Technol 60(1):243–246
20. Rajurkar KP, Kozak J (2001) Laser assisted electrochemical machining. Trans NAMRI 29:421–427
21. Poprawe R (Ed) (2011) Tailored light 2. Laser application technology. Springer, Berlin
22. Ha J-H, Lee Ch-M (2019) A study on the thermal effect by multi heat sources and machining characteristic of laser and induction assisted milling. Mater 12. https://doi.org/10.3390/ma12071032
23. Brecher Ch, Rosen C-J, Emonts M (2011) Laser-assisted milling of advanced materials. Phys Proc 5:259–272

24. M'Saoubi R, Axinte D, Leung Soo S, Nobel Ch, Attia H, Kappmeyer G, Engin S, Wei-Ming S (2015) High performance cutting of advanced aerospace alloys and composite materials. Ann CIRP Manuf Technol 64(2):557–580

Chapter 10
Mixed Hybrid Processes

10.1 Classification of Hybrid Machining Methods

Mixed or combined hybrid machining processes are basically classified within group (I) of the CIRP classification system of manufacturing processes [1, 2] which were overviewed in Chap. 2. However, it should be noticed that grouping of hybrid processes according to the rule that two or more constituent processes should occur more or less in the same time is not explicit and, in practice, can be related to the following cases [2–6]:

- Generally the hybrid effect is achieved by combining and intensifying mechanisms of the constituent processes which are used separately as conventional processes. The main aim is to avoid some typical disadvantages when they are used individually.
- Constituent processes should directly influence the process behaviour in order to accelerate material removal and modify positively physical phenomena in the machining zone.
- There are three general strategies, i.e. hybrid processes can be generated by a combination of subtractive machining processes, as for instance two conventional cutting processes with different kinematic features (turn-milling, turn-broaching), two non-conventional processes (EDM and ECM), one conventional and one non-conventional processes (combination of EDM or ECM with grinding), subtractive and transformative processes, as for instance cutting and burnishing processes, or only different transformative processes, i.e. deep drawing and cold forging.

In the second CIRP group (II) some hybrid processes with a controlled application of their mechanisms are selected, including two subtractive and transformative processes (as for instance grind-hardening or sequential cutting and burnishing) or two transformative processes (cryogenic deep rolling when the machining zone is cryogenically cooled by CO_2 snow) [1, 2]. These processes are presented in Chap. 12.

Currently, the hybridization of manufacturing processes is extended to tools and machine tools or multifunctional manufacturing platforms. In this case, hybrid

W. Grzesik and A. Ruszaj, *Hybrid Manufacturing Processes*, Springer Series in Advanced Manufacturing, https://doi.org/10.1007/978-3-030-77107-2_10

machine tools can be classified as sequential, assisted and combined. Sequential machine tools integrate constituent processes on one manufacturing platform, i.e. turn-milling machining centre which is capable of performing turning and milling operations (also hole making and grinding operations) in one workpiece clamping, i.e. two or more cutting processes with different kinematic features. It should be emphasised that hybrid processes cannot be simply associated with hybrid machine tools on which a number of various machining and non-machining processes are performed and they are not considered as constituent processes.

In general, assisted machine tools are additionally equipped with a laser source or an ultrasonic vibration generator (also the system for supplying cryogenic media) in order to perform different assisted processes described in Chaps. 6, 7 and 9. On the other hand, combined machine tools are mainly dedicated to hybrid unconventional processes, as for example *mechano-electrochemical milling* (MECM).

Hybrid tools are specially designed for one-pass machining of one or several surfaces, or finishing operation of one surface with rigorous demand to its quality [2]. They are classified as monolithic, indexable (exchangeable) and multi-functional ones. Monolithic cutting tools consist of two or more tools with different technological functionalities. In general, they are more complex and costly in comparison to standard/commercial tools and basically are utilized in series and massive production systems. In particular, they allow performing cutting and burnishing operations in one setup in various sequences.

10.2 Combination of Subtractive Conventional Processes

10.2.1 Turn-Milling Process

In general, turn-milling process is a form of milling process of complex surfaces in which the workpiece additionally rotates instead of its linear feed motion as in milling. In fact, it can be considered as a hybrid cutting process which integrates the kinematic features of two classical cutting processes—rotational motion of the tool (spindle) as in milling and rotational motion of the workpiece as in turning. As a result, complex eccentric shapes, which cannot be generated in separate processes, are created. It is a highly efficient machining process with reliable chip breaking, which is a serious technological problem in turning operations. Some typical applications of turn-milling process are as follows [1, 7]:

- Machining of outer cylindrical surfaces when the radial feed is superimposed on the milling cutter (Fig. 10.1a),
- Machining of eccentric shapes by simultaneous tool motion in two perpendicular directions, as for instance bearing journals in crankshafts and camshafts,
- Machining of conical surfaces using 5-axis CNC control. In general, special plunge cutters are needed when the tool is moved in more than two directions.

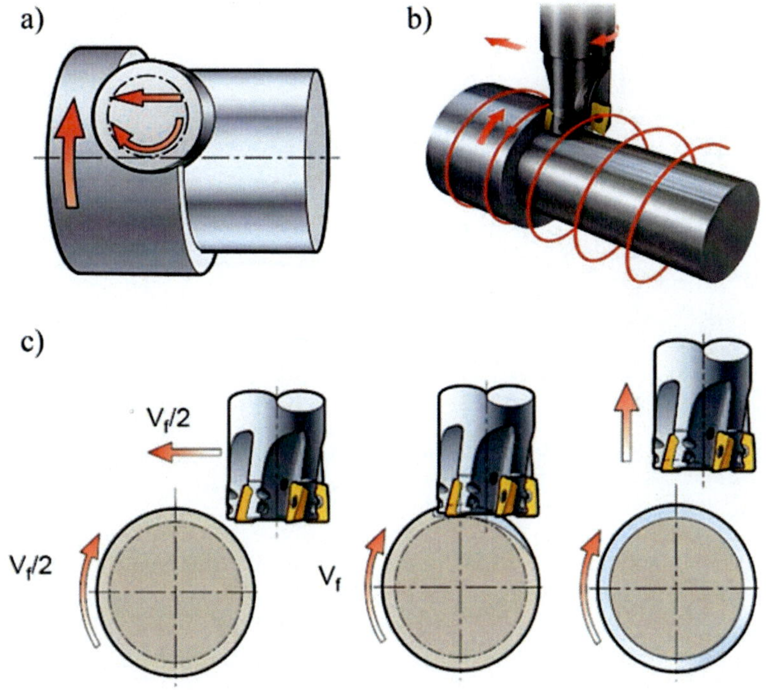

Fig. 10.1 Types of turn-milling operations: **a**, **b** face operations using special milling head or end-mill cutter, **c** different feed directions—face, plunge and tangential [7]

- Machining of 3D complex surfaces, as for instance turbine blades. In this case, 5-axis (or 4-axis) CNC control is used, i.e. 2 or 3 motions are performed by the workpiece and 1 or 2 by the tool.

In principle, turn-milling operations are the first choice for highly productive initial (rough) machining of complex-shaped parts (frequently non-rotating) on CNC lathe centres with additional use powered rotational tools. It is necessary to control movements along Y and C axes due to two simultaneous rotations of the workpiece and the milling cutter. Special milling heads and plunge (end-mill) cutters are used as shown in Fig. 10.2a, b. Some important advantages of this process cover high volumetric machining rate, the elimination of bending effect for long parts, longer tool life, the possibility of shaping eccentric parts and special shapes, i.e. polygonal parts (but additional movement in the X-axis direction is necessary). A serious limitation of this process is low geometric accuracy due to high shape errors resulting from visible regular scallops generated on the machine surface.

In the case of face milling, as shown in Fig. 10.1a, b linear contact between the tool edges and the workpiece allow producing quasi-cylindrical (regular polygonal) surfaces. In order to utilize effectively the tool in radial direction it is recommended to set it with at least two offsets (E_w in Fig. 10.2a), the first in the first tool rotation

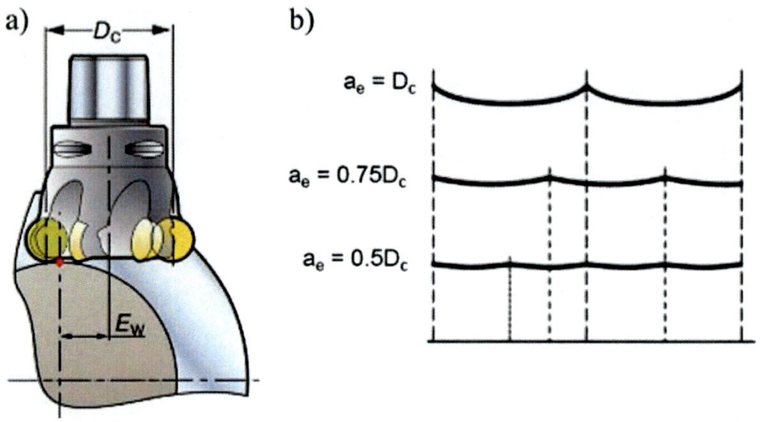

Fig. 10.2 Principle of tool offset in turn-milling (**a**) and selection of radial engagement ratio a_e/D_c (**b**) [7]

and the second in the next pass. Additionally, when the radial tool engagement (a_e) is higher than the cutter diameter, the tool should also be displaced in the axial direction, but not higher than 80% a_e per revolution [7]. The cutting width should be lower than 40% of the cutter diameter to reduce feed marks on the machined surface (Fig. 10.2b).

10.2.2 Turn-Broaching Process

In mass production of crankshafts a special machining operation termed rotational broaching, which can also be termed *turn-broaching* when considering special kinematic characteristics, is employed, As shown in Fig. 10.3a, b linear and rotational broachers are used. On the contrary, Fig. 10.4a, b present rotational inner and outer milling operations using special indexable and exchangeable disc-type milling heads with internal (IC) and external (EC) contact with the bearing journal surfaces. According to the production statistics from automotive industry the replacement of turning operations by rotational milling operations applied for crankshafts made of nodular iron allow the increase of tool life by 90% and the reduction of machining time by about 40% [1, 7].

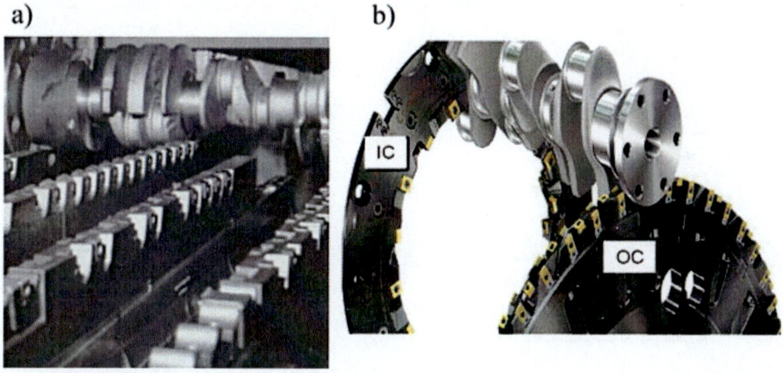

Fig. 10.3 Liner (a) and rotational (b) turn-broaching of crankshaft bearing journals using tools with internal (IC) and external (EC) contacts [1]

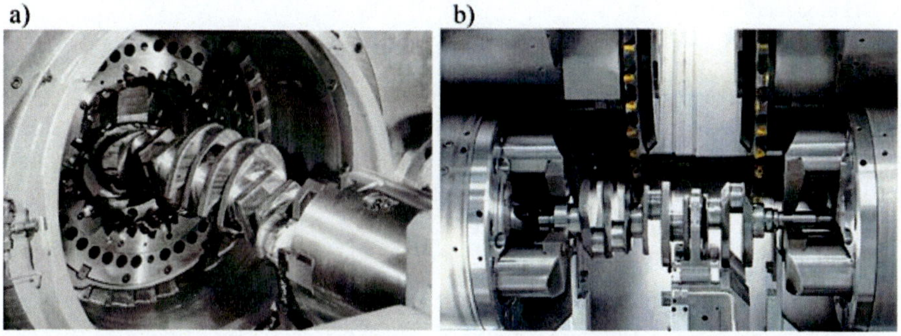

Fig. 10.4 Highly productive machining of crankshafts: **a** internal rotational milling, **b** external rotational milling of crankshaft bearing journal and face surfaces using a gang of disc-type milling cutters [1]

10.3 Combination of Subtractive Conventional and Non-conventional Processes

10.3.1 Electro-Discharge Grinding (EDG)

Hybrid electro-discharge (H-EDM) processes are developed to reduce the influence of the thermal action on the surface integrity, i.e. high surface roughness, micro-cracks, induced tensile residual stresses and alteration in micro-hardness. Mixed (compound) H-EDMs, EDM process can be hybridized with several processes such as electro-discharge grinding (EDG), abrasive electro-discharge grinding (AEDG), electrochemical discharge grinding (ECDG) and electrochemical discharge machining (ECDM). The ECDM process is characterized in Sect. 10.4.

In EDG process shown in Fig. 10.5a, conductive material is removed by spark

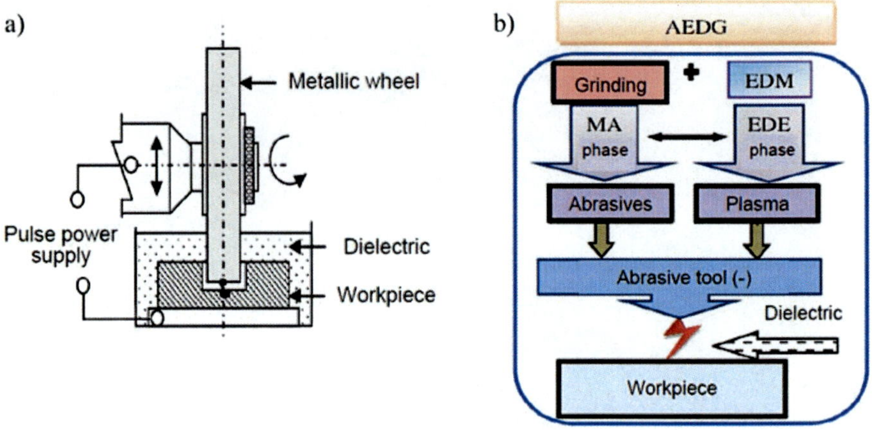

Fig. 10.5 Scheme of electro-discharge grinding (EDG) (**a**) and mechanism of material removal in AEDG process [8, 9]

discharges occurring between a rotating electrode tool and the workpiece which are separated by a flowing dielectric fluid. The rotary electrode wheel enhances flushing flow during the process so the molten particles are effectively ejected from the gap. In AEDG process, the metallic or graphite electrode used in EDG process is replaced by a metallic bonded grinding wheel. As a result (Fig. 10.5b) the material removal mechanism is based on synergistic interactive effect of the combination of EDM and grinding process. Therefore, the electro-discharge erosion (EDE) and mechanical abrasion (MA) effects occur. Electrical discharges thermally soften the material surface layer and reduce its mechanical strength, hence material removal is more energetically convenient. When erosion products are removed, new fresh workpiece surface is exposed and electrical discharges become more effective. It should also be noted that electrical discharges assist re-sharpening of abrasive grains in the wheel. According to many practical evidences the AEDG process is predominantly effective in machining superhard materials, engineering ceramics, sintered carbides and metal composites.

In AEDG process the surface layer properties can be comparable to the final result of conventional grinding or pure EDM process depending on the process variant shown in Fig. 10.6. In particular, as depicted in Fig. 10.6c, when the grinding in-feed is lower than the depth of discharge craters more rough surface is produced.

In general, the AEDG process is more efficient than conventional grinding process. Moreover, improved surface integrity, lower wheel wear and acceptable productivity are achieved. For instance, the detailed information on the machining of MMCs consisting of 60% Cu, 30% Fe and 10% C using AEDG process is documented in Ref. [10].

The abrasive-wire electro-discharge (AWEDM) process, shown in Fig. 10.7a, also uses the synergy between spark erosion and abrasion (Fig. 10.7c). In this process a metal bonded diamond abrasive wire is employed as presented in Fig. 10.7b. For

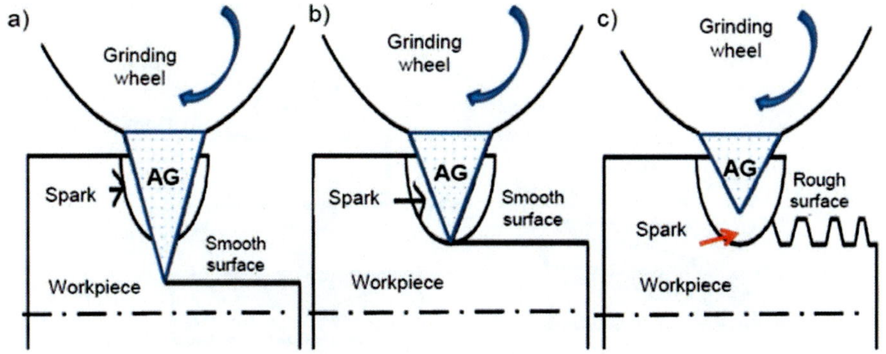

Fig. 10.6 Schemes of various variants of electro-discharge grinding process; **a** the machining depth is higher than depth of discharge craters; **b** machining depth is equal to depth of discharge craters; **c** machining depth is lower than depth of discharge craters [8, 9]. Symbol: AG—abrasive grain

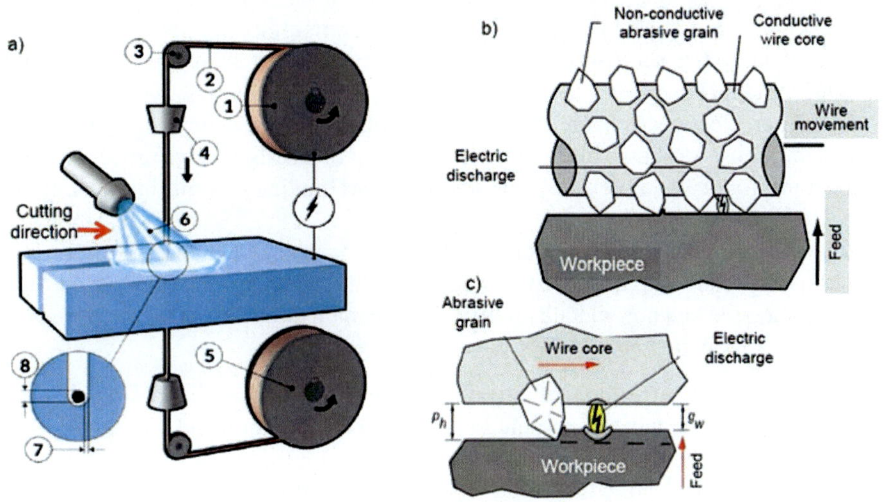

Fig. 10.7 Abrasive-wire electro-discharge cutting: **a** principle of the process, **b** movement of wire with bonded abrasives, **c** synergy between the grain abrasion and electric discharge [2, 8, 9]. Symbols: 1—wire feeder, 2—wire with bonded abrasives, 3—tension roll, 4—wire guide, 5—tension roll for worn wire, 6—dielectric jet, 7—working gap (kerf), 8—wire diameter

instance, during abrasive-wire EDM applied to cutting Ni-based alloys (Ni 600 alloy) such positive effects as higher process efficiency and better properties of the subsurface layer are reported [9, 11]. In particular, during AWEDM process with voltage of U = 108–162 V, impulse time of t_i = 4.9–100 μs, wire tension of 6.8 N, wire feed speed of 5.5 m/min, the volumetric material removal rate was equal to 0.02–0.10 mm^3/min and the gap width ranges between 8 and 20 μm. It revealed that the MRR is increased by several hundred percent in comparison to the WEDM process. The

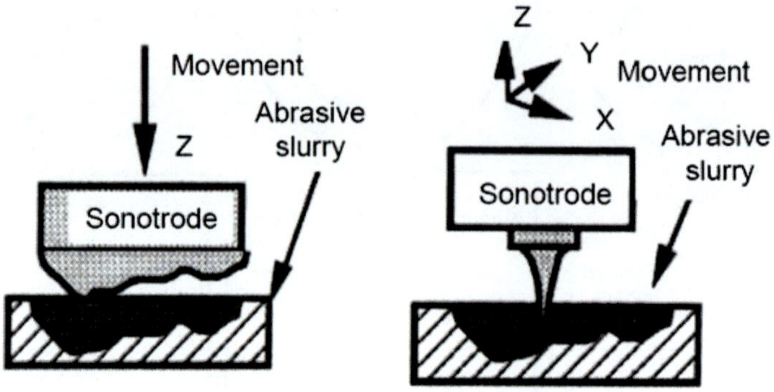

Fig. 10.8 Scheme of electro-discharge-ultrasonic-abrasive machining of complex surface using loose abrasives: **a** sinking operation, **b** milling operation

tension force acting in the abrasive bonded wire is only 1 N because abrasion action is performed on a thermally softened material. The surface roughness parameter Ra was equal to 0.9 μm (for comparison Ra = 2 μm when using the conventional WEDM variant [11].

Moreover, electro-discharge abrasive machining process can be assisted by ultrasonic vibrations. As shown in Fig. 10.8, this variant of EDM-abrasive process can be utilized to produce complex surfaces in both EDM sinking and milling operations (Fig. 10.8a, b respectively).

EDM-based hybrid processes enhance their performance which result in higher process efficiency and improved surface integrity when machining difficult-to-machine metallic and non-metallic materials. The main directions of further research include the following:

1. Mathematical modelling of physical phenomena in the channel of electro-discharges. The most important are electron concentration and plasma temperature.
2. Development of effective production techniques utilizing EDM-based hybrid manufacturing processes.
3. Development of multi-axis and ultra-precision machine tools for hybrid processes.
4. Enhancing monitoring techniques and metrological systems.
5. Exploring new hybrid processes.

10.3.2 Electrochemical Grinding (ECG) and Finishing

In electrochemical grinding (ECG) process, shown in Fig. 10.9a, a rotating metal-bonded wheel is made the cathode. In ECG material removal occurs as a result of electrochemical dissolution and abrasive action of micro-grains, i.e. material is

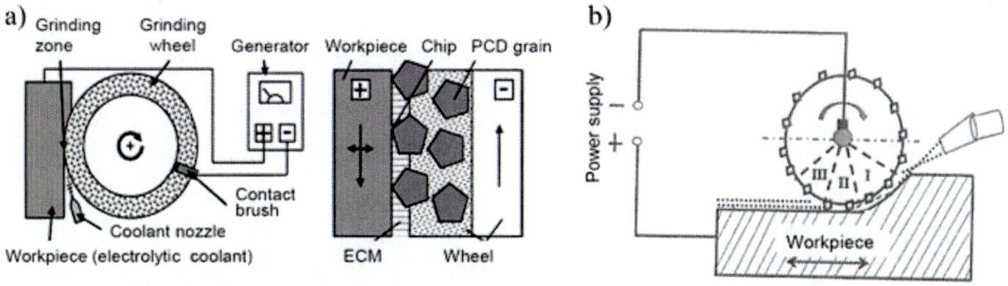

Fig. 10.9 Scheme od electrochemical grinding and interaction of abrasive grains on the workpiece material (**a**) and contact arc in ECG process (**b**) [2, 10, 12]

removed electrochemically and mechanically, and these two processes act synergistically. The electrochemical dissolution leads to formation of a passivation layer which is removed by abrasive action of the grinding wheel and a fresh workpiece surface is exposed every time. The electrolyte penetrates into small gaps (grinding zone in Fig. 10.9b) which are formed between the wheel and the workpiece. As shown in Fig. 10.9b the contact arc in ECG process can be divided into three zones. In zone I, electrolyte flow into the inter-electrode gap is assisted by the wheel rotation and material removal occurs at its leading edge. Gases produced in electrochemical reaction decrease the electrolyte conductivity and the inter-electrode gap becomes smaller. Thus, a small part of the workpiece material is removed as chips. In zone II, a passive film can be formed on the workpiece surface and the gas bubbles promote a more intensive removal rate. In addition, abrasive particles remove the non-reactive oxide layer. In zone III, material removal results completely from electrochemical dissolution and pressure is released slowly. It is important to note that in this zone burrs formed on the workpiece edges are removed.

On the other hand, micro-grinding is gentler than in classical grinding which leads to lower cutting forces and higher wheel durability. The fractions of electrochemical dissolution and abrasive action depend on the process parameters, including predominantly inter-electrode voltage, the load of the wheel to the workpiece (in case of elastic load) or the thickness of the removed material layer (in case of rigid load). The increase of the load or the thickness of the removed layer results in a lower inter-electrode gap width, which can cause local short-circuits or electric discharges. In turn, they additionally remove the material within the gaps between the abrasive grains and improve the grindability and durability of the grinding wheel [2, 10, 12].

Electrochemical abrasive machining can be utilized in the following main kinematic variants: electrochemical grinding (ECG), electrochemical machining using universal ball-ended wheel moving over 3D-shaped surface, and electrochemical machining with the assistance of powder abrasive particles. Figures 10.10 and 10.11 present some representative types of ECM grinding processes [12, 13]. It is important to notice that electric current is switched-off and the electrolyte supply is stopped in the final period of the machining process to improve surface finish to the level produced by conventional machining processes. ECG is well suited to machine hard

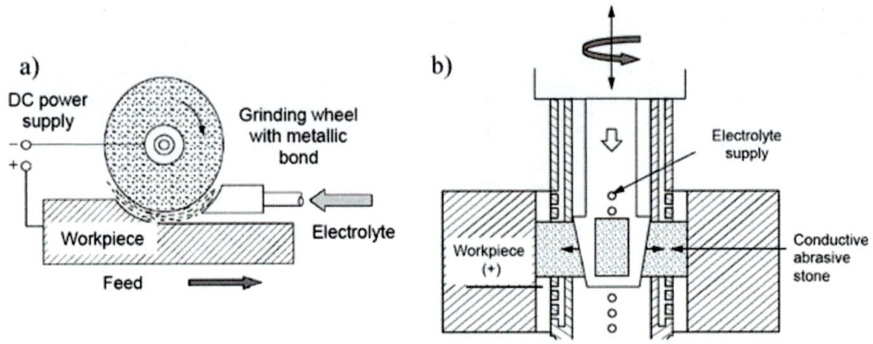

Fig. 10.10 Schemes of electrochemical grinding of a flat surface (**a**) and honing (ECH) of inner cylindrical surface (**b**) [13]

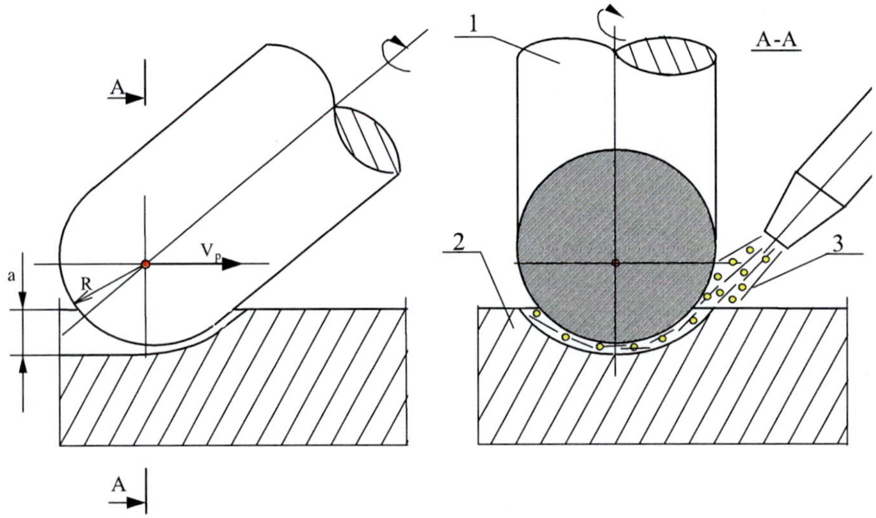

Fig. 10.11 Scheme of electrochemical-abrasive machining in electrolyte-powder mixture using special universal tool-electrode; 1—ball-ended tool-electrode 2—workpiece, 3—supplying of abrasive slurry

and fragile electrically conducted materials efficiently, economically and productively without affecting the functional properties of these materials. In particular, ECG is used for re-sharpening of cutting tools with brazed WC–Co cutting inserts. Other important applications include the burr-free grinding of honeycomb structures for turbine applications [3]. Within the combination of ECM and grinding, the electrolytic in-process dressing (ELID) also represents a major hybrid approach. It allows removing ECM-based anodic oxide layer on the grinding wheel [3].

Supplying a mixture of electrolyte and abrasive powder to the inter-electrode gap (Fig. 10.11) causes that basic material removal by dissolution is assisted by

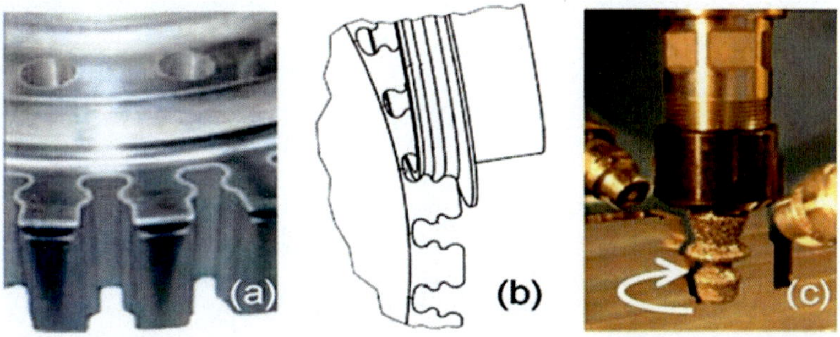

Fig. 10.12 ECM grinding of Christmas tree-type grooves in turbine disc after rough AWJM (**a**) cup wheel applied (**b**); (**c**) form wheel for ECM grinding [3]

additional micro-abrasive action which increases process efficiency and improves surface integrity. It is predominantly dedicated to finish machining of optical-quality (mirror) surfaces with extremely low surface roughness of $R_a << 0.1$ μm. It was revealed that after face grinding using loose abrasives with particle size of 2 μm the surface roughness was reduced markedly down to $R_a = 0.006$ μm (6 nm), and even to $R_a = 0.002$ μm (2 nm) when smaller abrasives of 0.5 μm size were used and machining time was equal to $t = 160$ s [2, 10].

It should be noted that one of the important applications of electrochemical grinding is finish machining of elements made from special materials such as sintered carbides, high-strength and heat-resistant alloys (Inconel, Nimonic), titanium alloys and metal matrix composites (for instance PCD–Co, Al–SiC, Al–Al_2O_3 grades). Depending on the material grade used, the obtainable surface finish is in the range of Ra << 0.1 μm. Specific areas of applications cover aircraft, automotive and medical industrial sectors. Such machining operations as micro-holes or 3D surfaces are widely performed [31]. For instance, as shown in Fig. 10.12, a Christmas tree-type grooves in the turbine hub for aircraft jet engines were pre-machined using AWJM cutting and they were finished using ECM grinding and a cup rotational electrode-wheel (Fig. 10.12b, c).

Figure 10.13 presents the concept of a sequential hybrid process which incorporates the ECDM process as a pre-machining operation (Fig. 10.13a) and PCD micro-grinding (RAG - *rotating abrasive grinding*) as a post-machining operation. It was proven to be effective in glass-type materials to produce complex miniature structures (Fig. 10.13c, d) [1]. The grinding process with PCD tools reduces the surface roughness generated in ECDM process from a few tens of a micrometre to about 0.05 μm in Ra parameter, as well as the total machining time. As shown in Fig. 10.13d, finish PCD grinding allows producing microstructures in a soda lime glass workpiece with sharp rims and edges and accurate dimples and cavities.

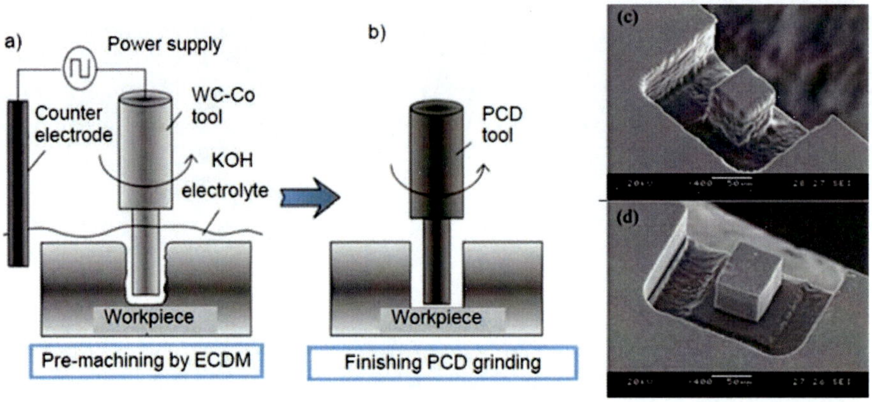

Fig. 10.13 Hybrid machining process combined ECDM process (**a**) and grinding using rotary pin with PCD grains (**b**) and examples of microstructures produced by ECDM (**c**) and PCD grinding (**d**) operations [1]

10.3.3 Electrochemical Finishing

One of the important functional goals of unconventional hybrid processes is to produce stress-free smooth surfaces. For instance, ECG process can be applied as a post-machining operation after rough EDM process of Ti-6Al-4 V alloy [9]. This is because the white layer is formed and micro-cracks occur on the machined surface. Such surface defects are not acceptable in aircraft industry because they substantially limit the fatigue strength of highly demanded elements. In addition, the obtainable surface roughness Ra ranges from 3 μm to about 6 μm depending on the current value. After ECG finishing the surface roughness is reduced to about 0.06 μm and white layer is completely removed.

Another proposal is electrochemical honing (ECH) which combines the capabilities of ECM with the capabilities of mechanical honing but at the same time it overcomes some limitations of these processes (mechanical honing suffers from reduced tool life, low productivity and inability to finish hardened parts). Therefore, ECH is used for finishing of internal cylinders and gears [2, 10]. It should be noted that ECH process is also termed mechano-electrochemical finishing because abrasives are not used [2, 10]. The ECH process is used for finishing of high-quality spur, helical and bevel gears. The main requirement is that the anodic workpiece gear meshes with a specially designed cathode gear to realize the mechanical honing. The honing gear is a helical gear and can be either an abrasive impregnated gear or made from material substantially harder than the workpiece gear. The principle of bevel gear finishing based on a twin complementary cathode gear is shown in Fig. 10.14.

As shown in Fig. 10.14, the workpiece gear (1), which acts as the anode, is mounted on the spindle of a bench drilling machine. On the other hand, a honing gear (2) is mounted on the rear of the workpiece gear. In the arrangement presented in Fig. 10.14 the cathode gears (3) and (4) meshes the workpiece gear (1). The

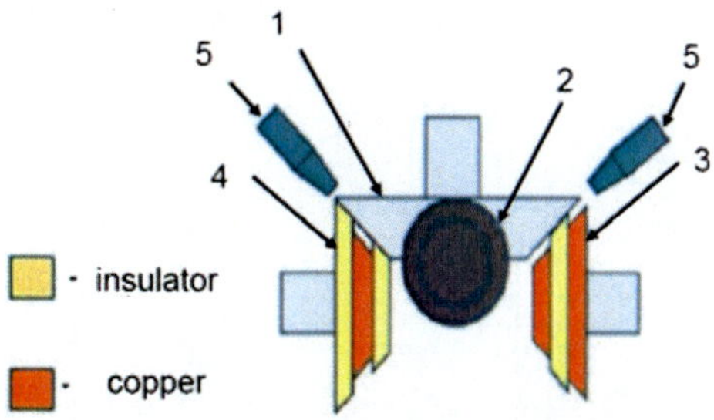

Fig. 10.14 Scheme of finishing of bevel gear using twin complementary cathode gears: 1—machined gear (anode), 2—honing gear-tool, 3 and 4—I and II gears (cathode), 5—nozzles supplying electrolyte to the working zone [2, 10]

cathode gear (4) has a conductive layer of copper which is sandwiched between two insulating layers of metalon, whereas the complimentary cathode gear (3) has an insulating layer of metalon between two conductive layers of copper. Electrolyte is supplied by two nozzles to the gap of 1 mm in width between the anode and cathode gears. During the honing process the electrolyte forms a passivating oxide layer on the workpiece gear tooth surface which inhibits further finishing. In the case study, gears (1) and (2) are made from 20 MnCr5 alloy steel of 50-54 HRC and 58-62 HRC hardness respectively, which is enough to abrade the passivating layer by the harder honing gear. After 2 minute meshing between the honing and workpiece gear the Ra parameter decreased down to about 1 μm (equivalently Rt parameter to about 8 μm). As a result, the accuracy of the workpiece gear is rapidly improved which causes lower gear wear and lower noise emission.

Figure 10.15 shows the principle of hybrid process combining electrochemical material removal and mechanical grinding for finish machining of precision small holes. As depicted in Fig. 10.15a spherical metal rod (electrically conductive) with nickel-coated diamond abrasives (electrically non-conductive) as cathode tool rotates at a high speed and simultaneously removes material both electrochemically and mechanically from a pre-machined pilot hole with diameter of D_0. In the process, material removal occurs in two phases (2) and (1). Phase (1) is entirely electrochemical action during which the passivation film is formed, whereas phase (2) combines electrochemical dissolution and mechanical grinding. The abrasive grains protrude the thin nickel layer and easily remove soft passivation layer and, thereby, phase 2 ends at the point where the maximum tool diameter is matched. Phase 3 is without removal action. For instance, this hybrid process was used for finishing holes of 0.6 mm in diameter and 0.3 mm in depth without burrs on the edges made in stainless steel parts [2, 4, 10]. They are made in fuel nozzles applied in aircraft engines. Minimum tolerance of the hole diameter of 0.12 mm, minimum surface roughness of Ra = 0.2 μm and zero conicity error were achieved for the rotation speed of 20,000 rev/min, voltage U = 4.5 V and axial feed rate of the tool of 5.5 μm/s.

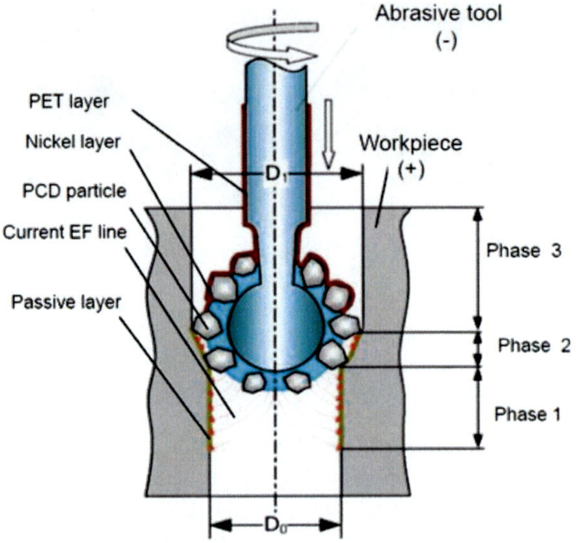

Fig. 10.15 Scheme of finishing micro-electrochemical-abrasive machining of initial hole of D_0 diameter; 1, 2 and 3—machining phases [2, 4]

10.4 Combination of Subtractive Non-conventional Processes

Ceramic materials and metal or ceramic matrix composites with appropriate chemical composition have very good mechanical properties (high tensile strength, fatigue strength, hardness and abrasive resistance) but they reduce the machinability rate when using conventional mechanical energy-based machining processes. In many cases machining processes are even not possible. The combination of ECM and EDM (ECDM) processes have been widely used to machine these materials (conductive and non-conductive) with extremely poor machinability rates [3, 9, 10, 14]. However, composites can be machined using both EDM and ECDM processes. As shown in Fig. 10.16a, the basic principle for ECDM is the discharge delay time of the EDM process, and, in consequence, electrochemically based surface dissolution is followed by local thermal material removal resulting from electrical discharges. The ECDM process involves two major phenomena: electrolysis leading to the generation of gas bubbles and arc discharge due to the breakdown of gas film. When the process parameters are properly selected smooth surfaces with reduced thermally affected zones and high geometrical accuracy of micro-structures are achieved.

A pulsed DC current is involved between cathode (tool) and anode. As shown in Fig. 10.16b the voltage (also current) impulse profile during the ECDM process consists of three phases [2, 3, 10]:

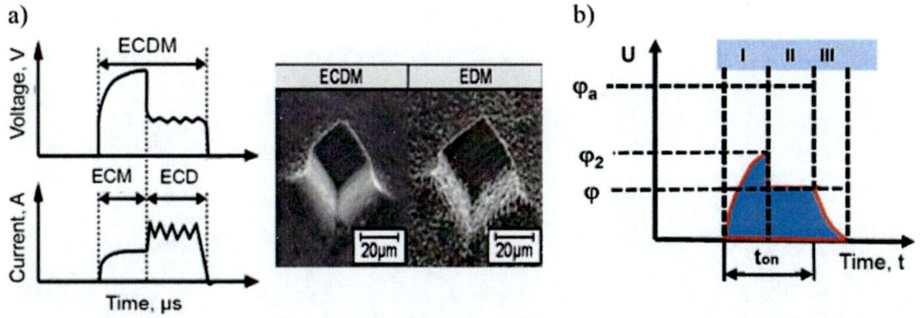

Fig. 10.16 Typical distributions of voltage and current impulses in ECDM process and application example (**a**) and characteristics of voltage impulse (**b**) [2, 3, 10]

- Phase I when inter-electrode voltage increases along with electrochemical dissolution of the metallic material phase and generation of ionic gas (hydrogen) bubbles up to achieving the critical voltage value (φ_2) and initiation of spark ignition. At this moment voltage drops down to the value of (φ),
- Phase II with the time duration of t_{on}, during which the spark erosion is continued to the end of voltage impulse (electric discharges commence at a critical voltage and current after the gas bubble insulating layer is broken down),
- Phase III during which voltage signal decreases continuously down to zero due to induction effect.

Glass, silicon and polymers are some of popular workpiece materials for MEMSs fabricating using ECDM process. Process conditions are selected to allow material removal by electrical discharges and due to intensive heating the temperature in the machining zone increases to 6000–12,000 K. As a result, material is removed by melting and vaporization, and sometimes by explosion due to excessive residual stresses [15].

The workpiece is immersed in the electrolytic solution (obviously water based solution of NaOH or KOH). DC voltage is supplied to the auxiliary electrode (anode) made from Pt and the tool (cathode to avoid anodic dissolution), which is located close to the non-electrically conductive workpiece, e.g. glass. The electrochemical reactions are as follows: on the anode—$4OH^- \rightarrow + O_2 + 4e^-$, and cathode- $2H_2O + 2e > 2OH^- + H_2$. When the voltage achieves the critical value, hydrogen bubbles form a film adjusted to the cathode and electrical discharges take place causing the gas film perforation. Because the electrode (cathode) is close to the workpiece, the material is removed during electrical sparks by melting and vaporization. The effective applications of ECDM micro-process were positively demonstrated in the machining of deep micro-holes and milling miniature complex structures in the range of 100 µm in Pyrex glass (Fig. 10.17).

It should be noted that in ECDM process electrochemical reactions are not involved in the material removal. They only support the electrical discharges, and electrical erosion is the basic mechanism of material removal. This is the reason for different naming ECDM process, i.e. electrochemical spark machining (ECSM) or

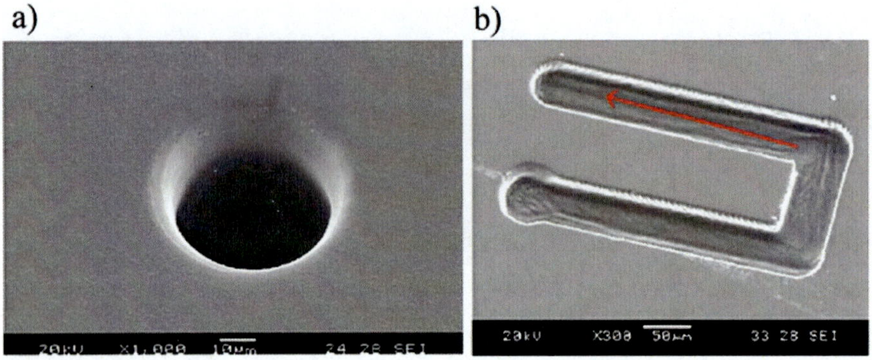

Fig. 10.17 Micro-hole (**a**) and a milled cavity (**b**) fabricated using ECDM in a glass plate (Pyrex glass grade) of 150 μm thickness [2, 15]

spark-assisted chemical engraving (SACE). The principle of the SACE process is illustrated in Fig. 10.18.

When a travelling wire is used as the tool electrode, i.e. cathode, the process is known as wire electrochemical discharge machining (WECDM) or wire spark assisted chemical engraving (WSACE). The process presented schematically in Fig. 10.19 is used for slicing brittle and hard, non-conductive materials. It is possible to make holes in Al_2O_3 ceramic parts (Fig. 10.20a) but for deeper holes the surface quality is not satisfactory due to micro-cracks resulting from the occuring thermal shocks. This negative effect can be partially reduced by replacing cylindrical metallic cathode by a metal-bonded abrasve tool (Fig. 10.20b). In this process, known as electrochemical discharge grinding (ECDG), a triple material removal mechanism incorporating mechanical abrasion, electric discharge erosion and electrochemical

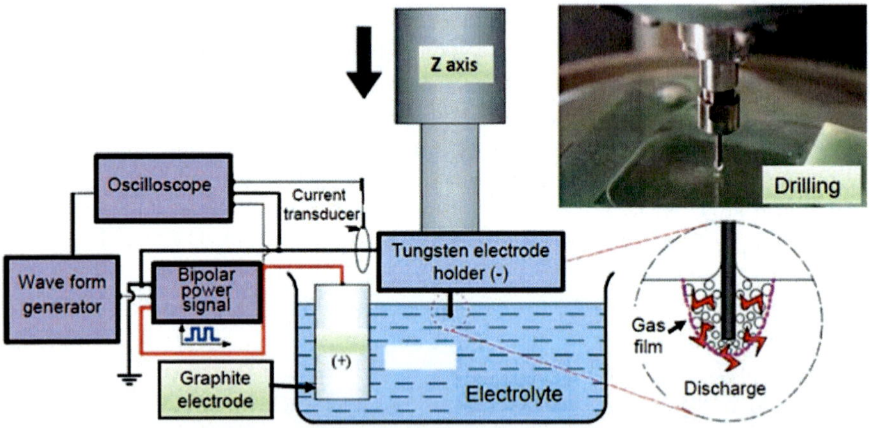

Fig. 10.18 Scheme of setup for testing spark assisted chemical engraving (SACE) process under sinking and drilling conditions [15, 16]

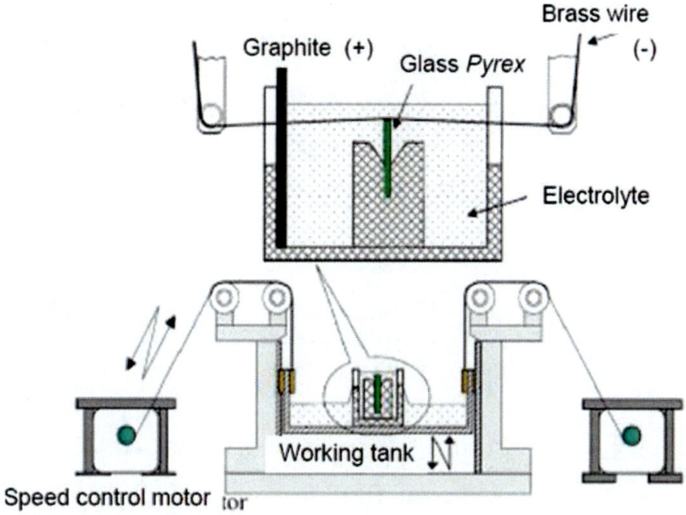

Fig. 10.19 Scheme of setup for wire cutting using SACE machining method [16]

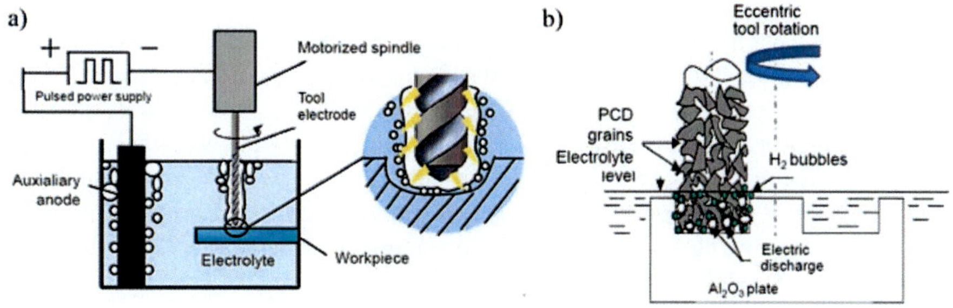

Fig. 10.20 Scheme of SACE machining process (**a**) and example of machining operation of Al_2O_3 ceramic element using PCD tool-electrode (**b**) [17, 18]

dissolution electrolytic erosion is developed. Abrasive coated rotary tool (cathode) serves as a grinding wheel which generate the sparks. In consequence, micro-holes in Al_2O_3, glass and metal matrix composite parts with a maximum depth of 1.5 mm and cylindrical surfaces by ECDM trepanning can be produced using abrasive electrode with a pulsed DC supply with higher efficiency and better surface finish.

It can be concluded based on the application examples provided that the ECDM process is the optimal choice for special conductive materials whereas the SACE process can be recommended for non-conductive materials.

Figure 10.21a presents the principle of a unique machining process which combines simultaneous micro-EDM and micro-ECM process (SEDCM-*simultaneous micro-EDM and micro-ECM milling*). Material removal is performed in low-resistivity deionized water layer- by-layer as a result of electrochemical reaction

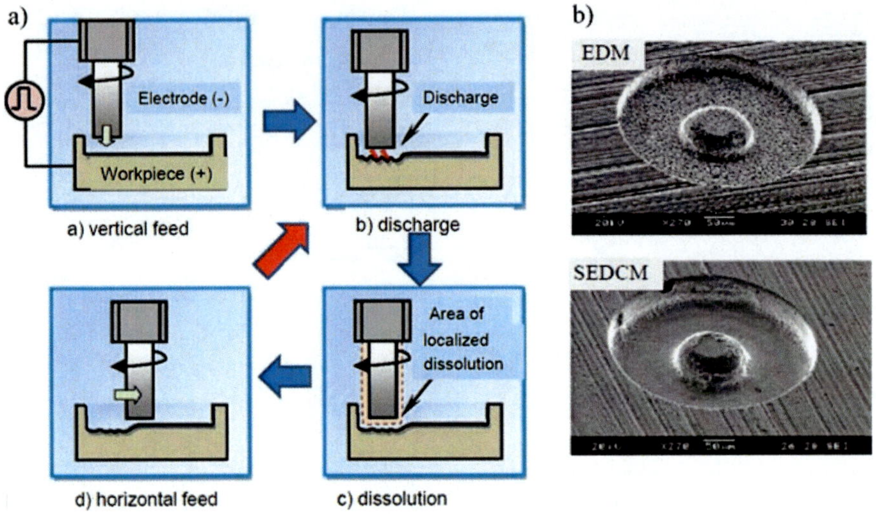

Fig. 10.21 Principle of simultaneous micro-EDM and micro-milling ECM (SEDCM) process (**a**) and examples of micro-cavities produced by EDM and SEDCM processes (**b**) [4, 18]

which occurs concurrently with electrical discharges. Figure 10.21b shows examples of micro-cavity with a central column produced by means of EDM and SEDCM milling. The work material was SUS 304 stainless steel and achievable surafce roughness was equal to 22 nm [4, 18]. However, it should be taken into account that intensive electrochemical erosion can induce surface damages because lower electrical conductivity is always observed in deionized water. Nanosecond voltage impulses and protection of the electrode by epoxide resin are recommended to apply in order to avoid excessive eletrochemical erosion.

References

1. Grzesik W (2018) Advances machining processes of metallic materials. Elsevier, Amsterdam
2. Luo X, Qin Y (2018) Hybrid machining. Academic Press (Esevier), London
3. Lauwers B, Klocke F, Klink A, Tekkaya AE, Neugebauer R, McIntosh D (2014) Hybrid processes in manufacturing. Ann CIRP Manuf Technol 63(2):561–583
4. Chavoshi S, Luo X (2015) Hybrid micro-machining processes: a review. Prec Eng 41:1–23
5. Zhu Z, Dhokia Y, Nassehi A, Newman S (2013) A review of hybrid manufacturing processes-state of the art and future perspectives. Int J Comp Integr Manuf 26(7):596–615
6. Unune D, Mali H (2015) Current status and applications of hybrid micro-machining processes: a review. Proc IMechEng J Eng Manuf 229(10):1681–1693
7. Turn-milling. www.sandvikcoromant.com
8. Satyarthi MK, Pandey PM (2013) Modelling of material removal rate in electric discharge grinding process. Int J Mach Tools Manuf 74:65–73
9. Ruszaj A, Skoczypiec S, Wyszyński D (2017) Recent development in hybrid manufacturing processes. Manag Prod Eng Rev 8(2):81–90

10. Gupta K, Jain N, Laubscher R (2016) Hybrid machining processes, perspectives and finishing. Springer, Heidelberg
11. Menzies I, Koshy P (2008) Assessment of abrasion-assisted material removal in wire EDM. Ann CIRP Manuf Technol 57(1):195–198
12. Bhowmik S, Divya Zindani D (2019) Hybrid micro-machining processes. Springer Briefs in Applied Sciences and Technology (eBook) https://doi.org/10.1007/978-3-030-13039-8 (Library of Congress Control Number: 2019931521)
13. El-Hofy H (2005) Advanced manufacturing processes. McGraw Hill, London
14. Liu JW, Yue TM, Guo ZN (2010) An analysis of the discharge mechanism in electrochemical discharge machining of particulate reinforced metal matrix composites. Int J Mach Tools Manuf 50:86–96
15. Cao XD, Kim BH, Chu HN (2009) Microstructuring of glass with features less than 100 μm by electrochemical discharge machining. Prec Eng 33:459–465
16. Zhan S, Zhao Y (2019) Suppression of cathode wear by a forward-bias voltage in pulsed spark assisted chemical engraving. Electroch Commun 111:1006643
17. Sanjay K, Chak P, Venkateswara R (2007) Trepanning of Al_2O_3 by electro- chemical discharge machining (ECDM) process using abrasive electrode with pulsed DC supply. Int J Mach Tools Manuf 47:2061–2070
18. Singh T, Dvivedi A (2016) Developments in electrochemical discharge machining: A review on electrochemical discharge machining, process variants and their hybrid methods. Int J Mach Tools Manuf 105:1–13

Chapter 11
Hybrid Processes with Controlled Mechanisms

11.1 Classification of Hybrid Machining Methods

In the second CIRP group (II) some hybrid processes with controlled application of their individual mechanisms are selected (Sect. 1.3). They cover two subtractive and transformative processes (as for instance grind-hardening or sequential cutting and sliding/roller burnishing) or two transformative processes (as for instance cryogenic deep rolling when the machining zone is cryogenically cooled by CO_2 snow) [1, 2]. These processes are characterized in detail in Sects. 11.3 and 11.4. In general, in group (II) the hybrid effect is achieved by combining and intensifying mechanisms of the constituent processes which are used separately as conventional processes and are typically performed on different devices after part transportation. A classical example is grinding and heat treatment to obtain the demanded material hardness. In a hybrid equivalent process, termed grind-hardening, material removal is combined with controlled in-air hardening by utilizing the heat induced directly during grinding process. The main aim is to obtain shorter technological chains and perform all technological operations in one setting and one manufacturing setup, which results in substantial economic and energy savings. Moreover, it allows avoiding some typical disadvantages when the constituent processes are used individually.

11.2 Synergetic Physical and Technological Effects

One of the principle hybrid methods of material shaping and generating demanded material properties is based on the broad applications of transformative processes which are performed with mass/volume change of the blank part [3–6]. They include material forming, surficial burnishing, heat treatment and some cryogenically-assisted processes. In conventional technological processes, they are performed individually as a sequence of entire machining/manufacturing processes. In order to shorten the technological chain and reduce process time and cost, as well as save

W. Grzesik and A. Ruszaj, *Hybrid Manufacturing Processes*, Springer Series in Advanced Manufacturing, https://doi.org/10.1007/978-3-030-77107-2_11

energy, they are integrated within one manufacturing platform. This approach cannot be achieved in individual constituent processes but in a hybrid process the mechanisms which are done in separate processes are controlled. It means the synergetic effect of physical/technological mechanisms is involved [2, 3]. The synergetic technological effect concerning turn-broaching and turn-milling machining processes was highlighted in Sect. 10.3. On the other hand, the synergetic physical effects, such as controlled phases of material microstructure leading to increased mechanical strength and hardness, obtainable by integrating subtractive and transformative machining processes, are overviewed in Sects. 11.3, 11.4 and 11.5 respectively.

11.3 Grind Hardening and Strengthening

Grinding and heat treatment which hardens the workpiece subsurface layer are frequently used in technological processes applied to highly functional parts with special demands with regard to dimensional accuracy, surface integrity and functional properties. This implies a great pressure to shorten technological chains by applying process hybridization (Sect. 1.3). One of the effective concepts is to replace conventional quenching of the workpiece surface by self-quenching during grinding process, termed *grind hardening* (GH). It utilizes the induced heat generated in the grinding process for local surface hardening on the workpiece. Examples of the applications of the GH process to plane and outer cylindrical surfaces are shown in Fig. 11.1a, b, respectively. In this hybrid process, controlling the thermal mechanism is based on the contact length (l_g) between the grinding wheel and the workpiece surface [7, 8]. Alongside the hardening effect also residual compressive stresses are induced to the subsurface layer.

During the grinding process the heat flux (Q_g) generated in the wheel-workpiece contact zone is transferred in a very short time to the subsurface layer causing its rapid quenching. For generating a high heat input rate (the thermal gradient), a grinding process with higher in-feed (as in plunge grinding) and slow feed speed to increase appropriately the heat treatment time is performed. The influence of the temperature gradient on the hardening depth (SHD) is shown in Fig. 11.2a. It is evident from Fig. 11.2a that higher SHD is achieved in peripheral face grinding than in cylindrical grinding operations. Obviously, higher SHD corresponds to lower temperature gradient independently of initial material microstructure. It is also important to note (Fig. 11.2b) that the GH guaranties more uniform profile of the Martens hardness (HM) in comparison to the induction hardening. In addition to the hardening effect, also microstructural changes and corresponding elastic deformation occur [3].

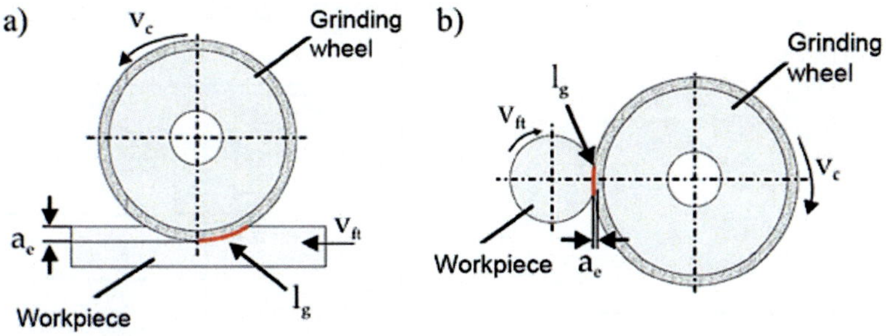

Fig. 11.1 Applications of grind-hardening to plane **a** and cylindrical **b** surfaces [8]

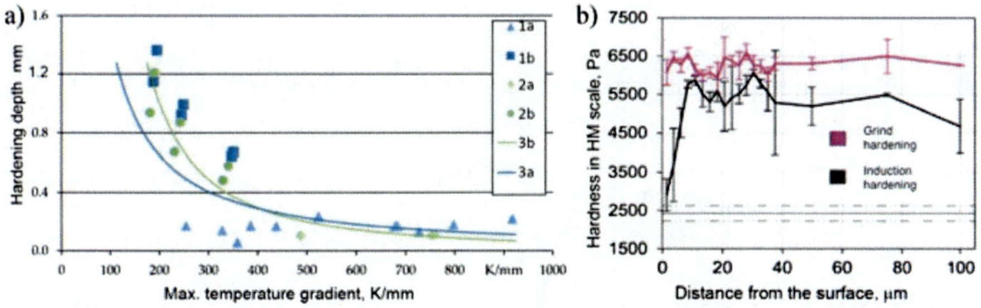

Fig. 11.2 Dependence of hardening depth on temperature gradient (a) and comparison of the effects of GH and induction hardening (b) [8, 9]. Symbols: 1a and 1b—cylindrical and face grinding of quenched and tempered material, 2a and 2b—cylindrical and face grinding of normalized material, 3a and 3b—appropriate trend lines

11.4 Combination of Rolling and Cryogenic Hardening

Hybrid cryogenically assisted machining in which LCO_2 cooled to the temperature of about −40 °C is injected during deep rolling is a relatively new technique used to maintain the effect of austenitic transformation carried out at 1145 °C during strain hardening [2, 10, 11]. The principle of the cryogenic deep rolling (*CDR—Cryogenic Deep Rolling*) from the point of view of the thermodynamic approach, which is based on the superposition of Gibbs energy changes caused by mechanical (ΔRht_{mech}) and thermal ($\Delta\ Rht_{therm}$) effects, is shown in Fig. 11.3a.

It should be noted in Fig. 11.4a that under three machining conditions—turning (T), ambient temperature deep rolling (DR) and cryogenic deep rolling (CDR)—the subsurface layer has a comparable hardness of approximately 700 HV. The result of the CDR technique is an increase in maximum hardness in the surface layer to 940 HV at the depth of 0.2–0.3 mm and generally at the depth of up to 1.5 mm (Fig. 11.4a), which results in inducing high compressive stresses up to about 800 MPa [11]. It should be noted that after turning the hardness of the core material is about 400 HV

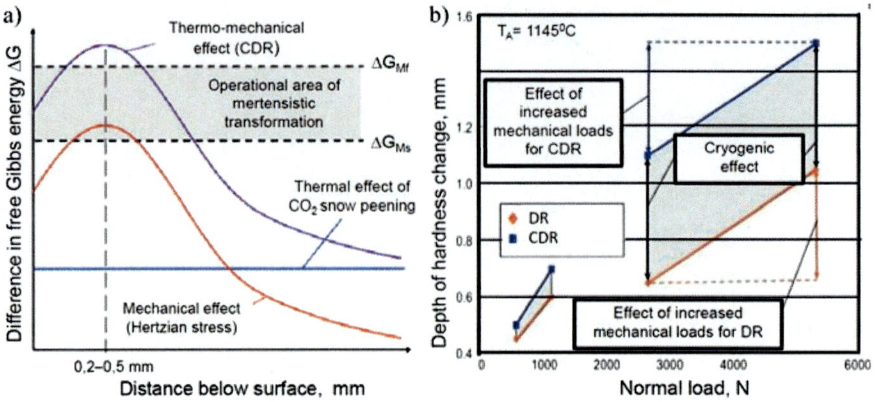

Fig. 11.3 Changes of Gibbs's energy resulting from cryogenic treatment of workpiece surface with CO_2 snow: **a** schematic diagram; **b** real changes in function of normal load [10, 11]

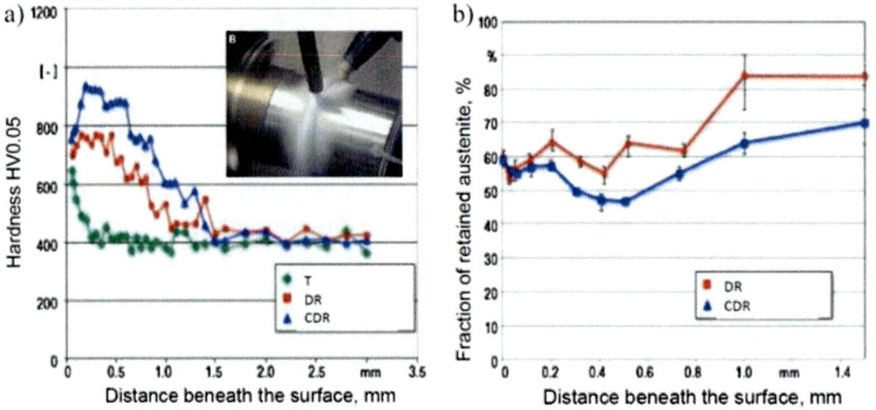

Fig. 11.4 Changes of hardness **a** and percentage of retained austenite **b** resulting from cryogenic treatment of workpiece surface with CO_2 snow. Workpiece material—AISI D3 (DIN X210Cr12 tool steel [2, 10, 11]

(Fig. 11.4a). Figure 11.4b confirms that the thermal conditions during the cryogenic cooling of the surface do not essentially affect the retained austenite (AS) content. Especially very close to the surface, at the depth of about 0.1 mm, the AS content for both cases decreases compared to the initial condition (about 80%).

11.5 Combination of Subtractive and Transformative Processes (Sequential Cutting and Burnishing Processes)

It is well known that sliding or roller burnishing as a subsequent pass after turning or milling, even grinding, operations improves substantially surface finish, reduces surface roughness, increase bearing ratio of the machined surface and induces high compressive residual stresses to the subsurface layer. This effects result from intensive plastic deformation of surface asperities and the machined surface becomes very smooth, frequently of a high optical reflectivity, i.e. optical quality. Burnishing as a finishing and/or strain-hardening machining process can be successfully applied to differently-shaped surfaces, from plane and cylindrical to complex 3D sculptured surfaces (Fig. 11.5). It is possible to burnish outer and inner surfaces using special burnishing tools.

As mentioned above, depending on the form of machined surface, the integration of sequential cutting and burnishing operations is realized on a CNC lathe (Fig. 11.6a) or a CNC milling centre (Fig. 11.6b). In the case of the CNC lathe cutting and burnishing tools are mounted in the turret and their use is controlled by the CNC program. Hence, these two subtractive and transformative processes are integrated based on the technological synergy. When the milling and burnishing processes are performed sequentially the main goal is to remove the feed marks (scallops) remaining on the complex surface by a ball-end milling cutter (Fig. 11.6b) using different strategies, i.e. the motion directions can be parallel or perpendicular. In

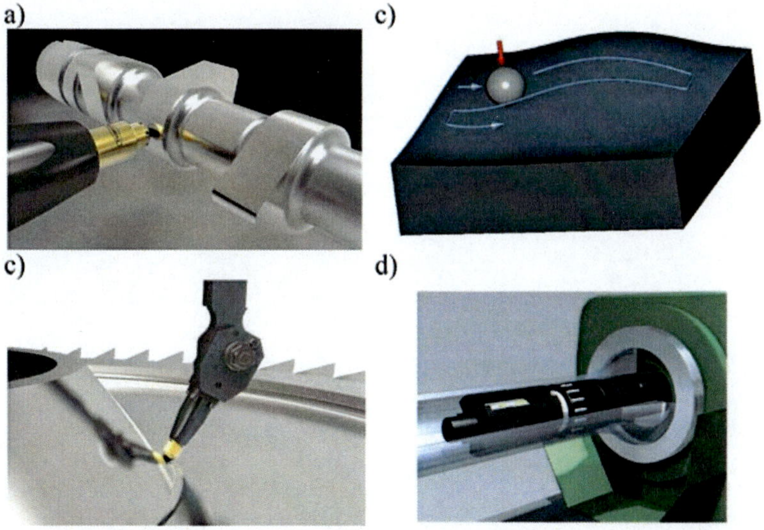

Fig. 11.5 Examples of burnishing processes: **a** cam surface of a cam shaft, **b** sculptured 3D surface, **c** a steam turbine blade, **d** a hole in a hydraulic cylinder after skive-burnishing [12]

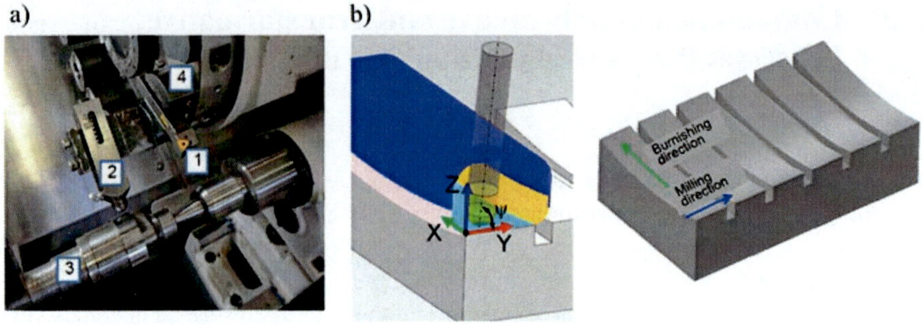

Fig. 11.6 Integration of cutting and finish burnishing processes on CNC machine tools: **a** 2/3-axis CNC lathe, **b** 3-axis CNC milling [12, 13]

Fig. 11.6b the directions of feed motions for ball-end milling and sliding burnishing are perpendicular (crossing), as for instance in finishing concave or convex surfaces on dies or moulds instead of costly mechanical polishing [13].

Figure 11.7 presents some important results concerning sequential turning and ball burnishing process regarding surface topography (Fig. 11.7a) and modification of the bearing area (material ratio) curve (BAC) (Fig. 11.7b) by means of different machining strategies, i.e. one or multi-pass machining. It can be seen that BACs denoted by numbers 1, 2 and 3 are determined after turning operations, whereas BACs with numbers 4, 5 and 6 are determined after burnishing using different process conditions (variable normal force and feed rate) and number of passes, e.g. curve #5b was obtained after two-pass burnishing. The distributions of BAC curves in Fig. 11.7b

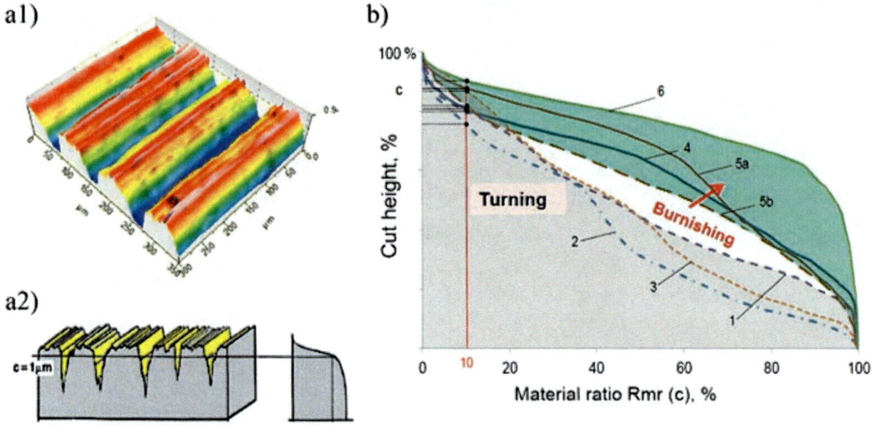

Fig. 11.7 Effect of hybrid process including turning and burnishing operations in relation to surface topography **a** bearing ratio **b** for 41Cr alloyed steel hardened to about 55 HRC [14]

reveals the substantial increase of material/bearing ratio for a *plateau*-type surface produced (Fig. 11.7a2). In particular, the maximum roughness height (Rt) decreases by 20–50% depending on process conditions employed [14]. As a result, fatigue strength and the resistance to contact loads are improved.

11.6 Industrial Applications of Machining Processes with Controlled Mechanisms

As mentioned in Sects. 11.3–11.5 hybrid processes which effectively utilize the controlled combinations of mechanisms of two separate processes have many industrial applications. Figure 11.8 compares the technological chains for conventional and grind-hardening based hybrid processes. It is evident that the hybrid process variant results in substantial time and cost savings due to reduced transportation and elimination of separate heat treatment process. The hybrid GH process is used in automotive industry in manufacturing geared shafts for manual gear boxes.

Figure 11.9a presents a triple diagram in which different machining operations were selected taking into account their possible impacts, mechanical, thermal and chemical, on the surface and sublayer properties. It can be seen in Fig. 11.9a that deep rolling and grind strengthening (top corner) exert primarily mechanical loads on the machined surface. Hard turning and grinding (also grind-hardening) are performed with additional thermal impact. However, grind-hardening requires more thermal influence than the classical grinding operations. In contrast, pure thermal loads are generated in both pulsed (μs) laser machining and induction hardening processes. The comparison of thermal loads generated in grinding and heat induction are illustrated in Fig. 11.9b. Planar and volumetric heat sources with different heat dissipation

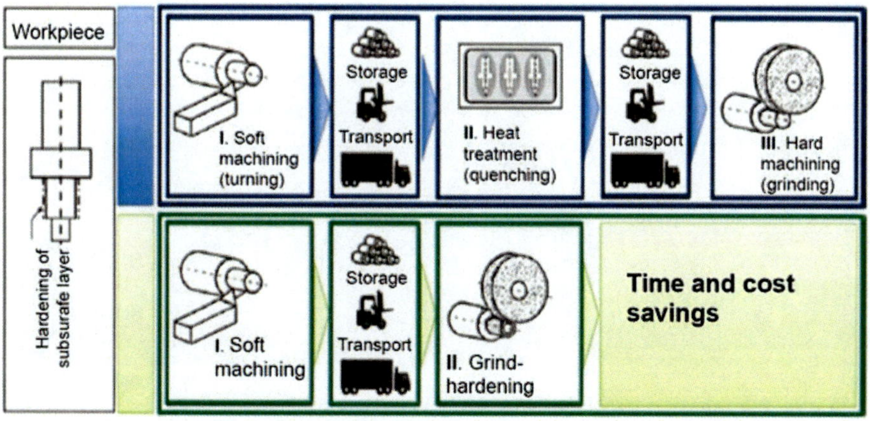

Fig. 11.8 Schematic illustrations of modifications of technological processes' structures when using grind-hardening operations [15]

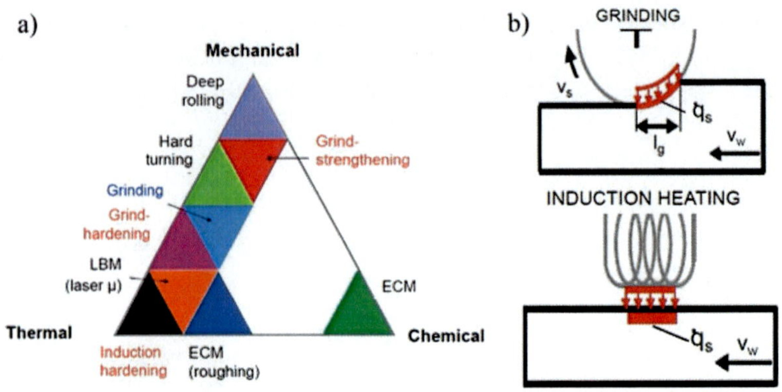

Fig. 11.9 Triple diagram showing integration possibilities for grinding and heat treatment, and burnishing processes (**a**) and comparison of thermal loads in grinding and induction heating (**b**) [9]

rates are generated, respectively. Similarly, EDM process is mainly characterized by the thermal load but for rough EDM also slight chemical influences appear. The process mechanism in ECM (lower right corner) is purely chemical (electrochemical). Because EDM and ECM processes are thermally and chemically homogeneous, they can be assisted and combined by a number of other processes, including abrasive, vibration, laser, magnetic field and electrodischarge and/or electrochemical effects.

As mentioned above, a visible trend that multi-axis machining centres are equipped with different burnishing tools and devices is observed, as shown in Fig. 11.10a. In general, the mechanical load is exerted hydrostatically by applying hydraulic pressure on the burnishing element and it is controlled by a CNC programming system (Fig. 11.10b). As a result, sequential processes are performed more

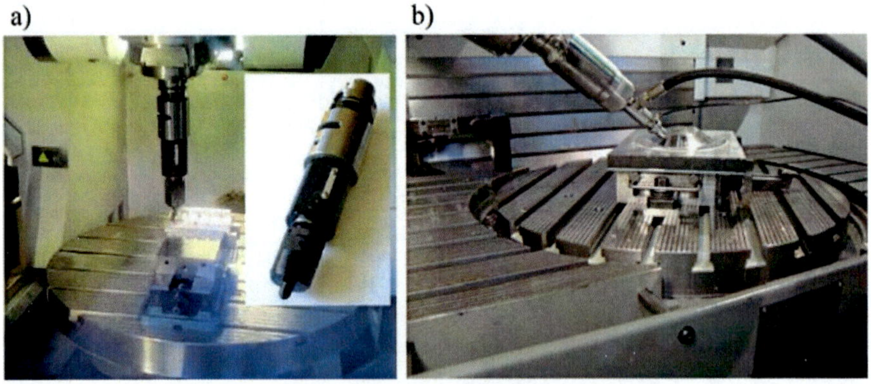

Fig. 11.10 Burnishing of 3D surfaces in 5-axis controlled machining centres: **a** vertical centre, **b** horizontal centre [13, 16]

effectively and better part functionality can be achieved. Moreover, burnishing operations can be carried out on free surfaces, predominantly on dies and moulds, turbine blades and elements with high fatigue life, e.g. transient surfaces on powered shafts.

References

1. Grzesik W (2018) Advanced machining processes of metallic materials. Elsevier, Amsterdam
2. Luo X, Qin Y (2018) Hybrid machining. Academic Press (Elsevier), London
3. Lauwers B, Klocke F, Klink A, Tekkaya AE, Neugebauer R, McIntosh D (2014) Hybrid processes in manufacturing. Ann CIRP Manuf Technol 63(2):561–583
4. Chavoshi S, Luo X (2015) Hybrid micro-machining processes: a review. Prec Eng 41:1–23
5. Zhu Z, Dhokia Y, Nassehi A, Newman S (2013) A review of hybrid manufacturing processes-state of the art and future perspectives. Int J Comp Integr Manuf 26(7):596–615
6. Unune D, Mali H (2015) Current status and applications of hybrid micro-machining processes: a review. ProcIME J Eng Manuf 229(10):1681–1693
7. Kolkwitz B, Foeckerer T, Heinzel C, Zaeh M, Brinksmeier E (2011) Experimental and numerical analysis of the surface integrity resulting from outer-diameter grind hardening. Proc Eng 19:222–227
8. Kolkwitz B, Kohls E, Carsten Heinzel C, Brinksmeier E (2018) Correlations between thermal loads during grind-hardening and material modifications using the concept of process signatures. J Manuf Mater Proc 2:20
9. Borchers F, Clausen B, Eckert S, Ehle L, Epp J et al (2020) Comparison of different manufacturing processes of AISI 4140 steel with regard to surface modification and its influencing depth. Metals 10:895
10. Mayer D (2012) Cryogenic deep rolling—An energy based approach to enhanced surface hardening. Ann CIRP Manuf Technol 61(1):543–546
11. Mayer D, Brinksmeier E, Hoffman F (2011) Surface hardening by cryogenic deep rolling. Proc Eng 19:258–263
12. Chudy R, Grzesik W (2016) Investigation of sequential machining processes in terms of power consumption and surface quality. J Mach Eng 16(3):67–74
13. Kalisz J, Grzesik W, Żak K, Czechowski K (2017) Properties of the subsurface layer after rolling burnishing of initially milled aluminium alloy. J Mach Eng 17(3):66–74
14. Grzesik W, Żak K (2014) Characterization of surface integrity produced by sequential dry hard turning and ball burnishing operations. Trans ASME J Manuf Sci Eng 136:031017-1–031017–9
15. Salonitis K (2006) Energy efficiency and environmental impact implications of grind-hardening process. In: Grind hardening process. Springer Briefs in Applied Science and Technology Berlin, pp 81–91
16. Grochała D, Berczyński S, Grządziel Z (2014) Stress in the surface layer of objects burnished after milling. J Adv Manuf Technol 72:1655–1663

Chapter 12
Hybrid Additive and Subtractive Processes

12.1 Applications of Different Additive Technologies and Subtractive Processes

12.1.1 Introduction

The realisation of the Manufacturing 4.0 strategy is closely linked to *the hybridisation of* manufacturing processes, as signalled in Sect. 1.4. The scope now also includes hybrid processes and equipment that combine *Layered/Additive Machining* (LM/AM) and Subtractive Machining (SM) CNC (LM/SM CNC) which are not included in the basic classification of hybrid machining processes [1]. Typical three applications are the manufacturing of parts entirely from powder materials or only some complex elements, the application of layers from materials other than the base material and the repair/restoration of worn parts.

Hybrid machining, in its variant that combines first the manufacturing of the part by additive technology and secondly the finishing of the part by subtractive machining (mainly 3D milling), is a compromise between the high complexity of the product design characteristic for additive machining and the high productivity achievable with CNC machining. It should be noted that the productivity of the AM process is an order of magnitude lower than that of CNC machining and users always have the dilemma of choosing either higher productivity (greater thickness of the material layer applied) or better quality (lower thickness of the material layer). The main technological problem that occurs in additive machining is surface unevenness (staircase error) that depends on the thickness of the material layers deposited (Fig. 12.1). As a result, the hybrid AM/CNC manufacturing process becomes profitable for smaller and medium-sized batches because of the possibility of independent control of productivity and surface quality [1, 2].

The range of possible solutions is very wide, e.g. it is possible to shape additional elements of an object by 3D printing and *laser metal deposition* (LMD), to apply

W. Grzesik and A. Ruszaj, *Hybrid Manufacturing Processes*, Springer Series in Advanced Manufacturing, https://doi.org/10.1007/978-3-030-77107-2_12

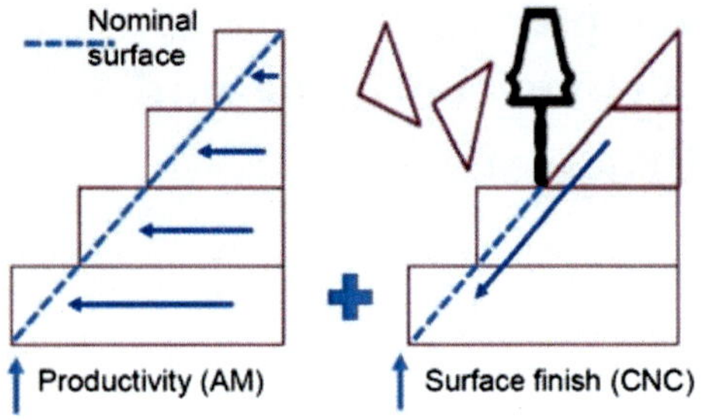

Fig. 12.1 Principle of integration of AM and CNC machining processes in a hybrid process (AM + CNC) [3]

elements made of other materials than the base material, but also to repair expensive or damaged/worn elements.

The combination of these complementary techniques is now standard practice applied to most metal products manufactured by AM techniques when surface quality (functionality), narrowed dimensional tolerances, avoidance of the risk of excessive stresses, or even aesthetic considerations are required. An important factor is the drastic reduction in waste that results from using thin sheets or powders instead of monolithic semi-finished products. This makes it possible to minimise the material removal in finish machining. It should be added that up to 80–90% of the volume of the bulk material is removed when shaping complex monolithic components from aluminium and titanium alloys in aerospace industry [2].

12.1.2 Layer Deposition Techniques in Additive Processes

The industrially used additive machining techniques are discussed in Sect. 1.2. In Sect. 12.1, different methods of making a semi-finished product by additive processes are presented together with the subsequent machining operations. Figure 12.2 shows the possible machining variants in the first and second phase of the hybrid manufacturing process. Powder Bed Fusion (PBF) and Direct Energy Deposition (DED), which also includes the very popular Laser Metal Deposition (LMD) process, are included as additive shaping methods. It should be noted that the DED process is replacing the previously frequently used selective laser sintering (SLS) process due to its higher productivity, but also the lack of a chamber and support structures [4]. In the LMD process, the typical material deposition rating is 5–30 g/min, while in the SLM process it is only 2–3 g/min.

Figure 12.3 shows a hybrid manufacturing process (A/SM) consisting of selective laser melting (SLM) and CNC milling. The digital CAD model is divided into a number of thin layers, which are then virtually successively deposited using selective laser melting. The base material (substrate) is clamped to the machine table. A part is created from the powder as the laser scans the surface and melts it. The process is repeated and progresses as the platform is lowered one layer at a time and the applied powder layers are successively melted to form a fragment of the sintered part (2),

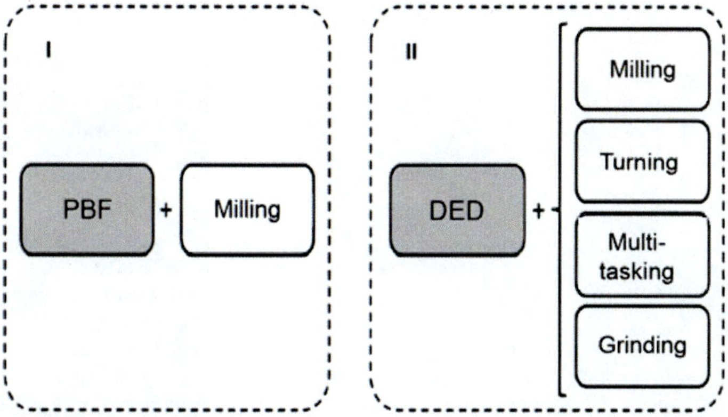

Fig. 12.2 Different combinations of additive and subtractive processes in existing hybrid machine tools [4]. Symbols: I—PBF based AM, II—DED based AM

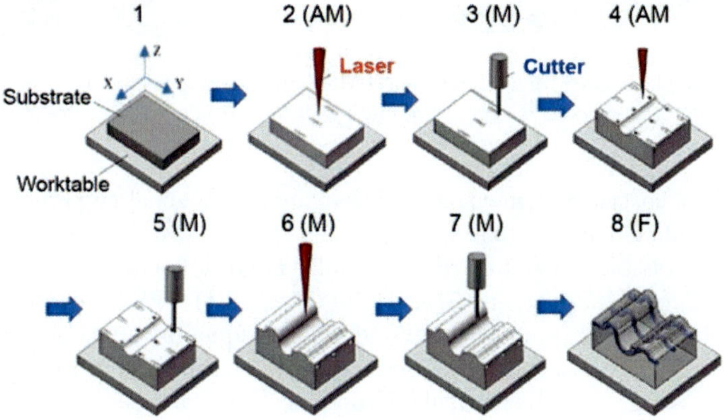

Fig. 12.3 Scheme of a hybrid machining process A/SM [6]. Symbols: AM—additive machining, M—milling, F—finishing machining (obviously polishing)

which is then milled (3). The complete forming of the part follows after two more stages of alternating additive and subtractive machining (4) and (5), and (6) and (7) respectively). It should be noted that milling usually removes dimensional and shape errors introduced in the previous AM operation. Complex internal profiles, such as cooling channels, can be produced.

The principle of laser cladding used in this AM process is illustrated in Fig. 12.4a. Additive machining by laser cladding is carried out using deposition heads of different capacities, e.g. the basic HS LMD head (*high speed LMD*), while small additional parts, e.g. fins or pins, are formed with the F LMD head (*fine* LMD) (Fig. 12.4b). What may be surprising is that the total part manufacturing time is not shorter than that of a monolithic part, however, the decisive factors are much lower material

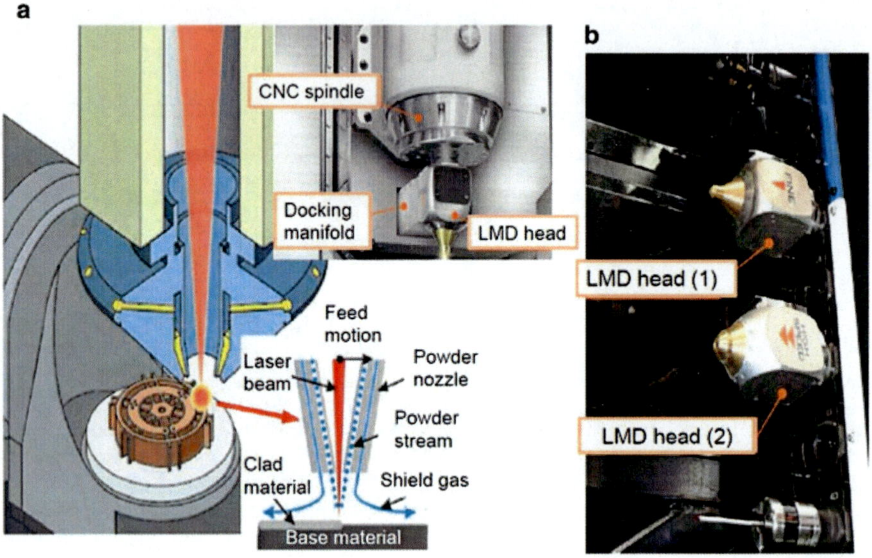

Fig. 12.4 Principle of Laser Metal Deposition (LMD) process (**a**) heads for layer deposition with different efficiency mounted in the spindle of milling machine (**b**) [7, 8] (www.mazak.co.jp)

usage and tool costs and maintenance of the required functional properties (thermal, corrosion, oxidation and creep resistance).

A new idea is the integration of the popular rapid prototyping technique FDM (*fused deposition modelling*) and CNC milling (Fig. 12.5), which is especially effective for polymer parts printed without additional supports or reinforcement with metal infills [9]. CNC machining is used for finishing, trimming or shaping of free-form surfaces (Fig. 12.6). Its particular importance consists in removing the adverse effects of the FDM process on the shape of the component due to high material

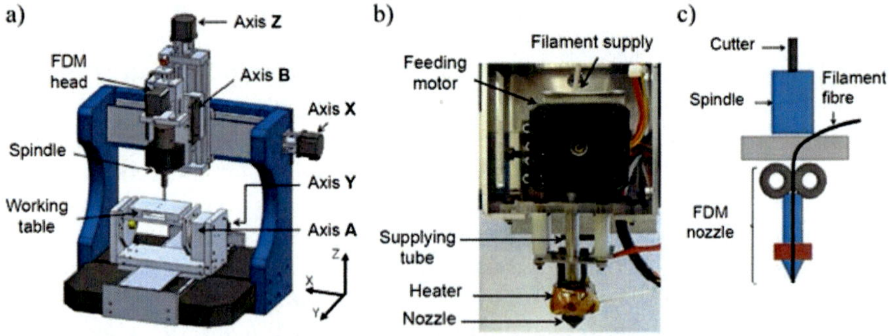

Fig. 12.5 Design of hybrid machine tool integrating FDM and CNC machining processes: **a** scheme of the machine structure, **b** design of FDM extruder, **c** working principle of extruding head [9]

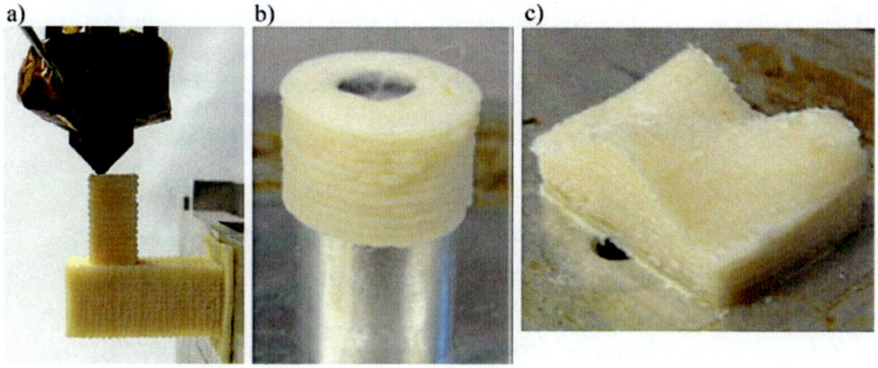

Fig. 12.6 Examples of parts machined on a hybrid (FDM + CNC) machine: **a** part with overhang (Γtype), **b** stepped element made from aluminium embedded in polymer, **c** element with free-form surface after milling [9]

shrinkage and excessive residual stresses. The design of a prototype hybrid machine tool equipped with a head for liquid material extrusion is shown in Fig. 12.5a, b (the principle of the extruding head operation is presented in Fig. 12.5c).

Figure 12.7 shows a technological chain for the hybrid AIMS (*Additive system Integrated with subtractive MethodS*) process, which involves rapid prototyping using the EBM method and subsequent rapid CNC machining after clamping using previously made lateral sacrificial supports. The AIMS process requires two types of sacrificial supports necessary to achieve overhang during layer deposition and to fix

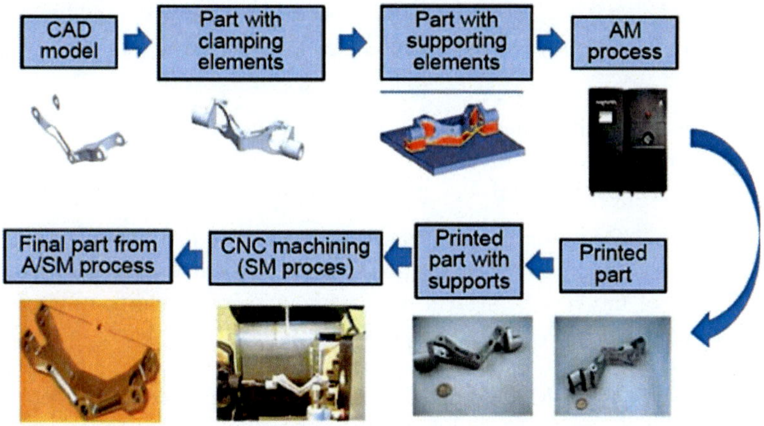

Fig. 12.7 Flow of the AIMS hybrid process including AM (EBM—*electron beam melting* or DMLS) and SM (CNC-RP) operations [5]

the part in the clamp of the CNC machine tool. It should be added that the tool change time is not included in the total manufacturing time. Therefore, the total fabrication time can increase as a result of lower tool life and more tool changes, which is easily noticeable for the hard-to-machine titanium alloy Ti-6Al-4V [5].

12.2 Special Multi-axis Manufacturing Platforms for Hybrid Additive–Subtractive Processes

12.2.1 Constructions of Integrated Modular Manufacturing Platforms

In the case of AM/SM hybrid processes, the integration/combination of several processes in a single manufacturing machine tool, called a hybrid machine tool, has been developed on the basis of multifunctional CNC machine tools for integrated (complete) machining (termed *multi-tasking machine tools*), such as multi-axis milling centres or turn-milling centres [3, 4]. The first stage of integration, dating from the late 1980s and early 1990s, involved the installation of a laser source to assist thermally the conventional machining process. The current stage of hybridisation of manufacturing/machining processes concerns the implementation of additive machining (AM-additive) and finish/machining (SM-subtractive) with automatic dimensional control in a single machine [1, 4]. Additive machining is usually referred to as "3D printing".

Previous research on a laboratory and industrial scale have been directed towards the creation of a modular manufacturing platform integrating various additive manufacturing (AM) techniques with subtractive machining (SM CNC), e.g. [3, 5, 10, 11]:

- FDM/extrusion in different variants of CNC machine tools,
- SLM on VMC vertical centres,
- Direct metal deposition (DMD), e.g. laser cladding on 5-axis machining centres.

The main design problem in the synergistic integration of AM and CNC techniques is the installation of the deposition system on new or used CNC machine tool regardless of the control system used without special modifications or after appropriate *retrofitting*. A pioneering research project completed in 2008 was RECLAIM (**RE**manufacture of high value products using a **C**ombined **LA**ser cladding, **I**nspection and **M**achining system) [3, 12], which recommended a cost-effective upgrade of new CNC machine tools by installing a laser deposition module (Fig. 12.8). It should be mentioned that one of the first hybrid machine tools integrating laser cladding/deposition on a CNC milling machine was used already in 2004 for manufacturing materials/products for the aerospace industry in the production of the F-15 fighter aircraft [13].

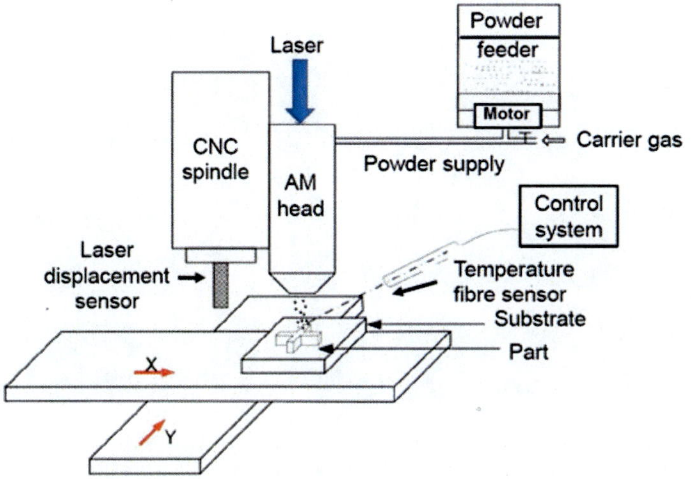

Fig. 12.8 Structural scheme of prototyping machine tool with milling spindle and laser deposition head [13]

A scheme of the design and operation of a 5-axis prototype hybrid machine tool that was designed at the Laser Assisted Manufacturing Processes Laboratory (LAMP) at the University of Missouri in the late 1990s is shown in Fig. 12.8. The machine tool is based on the integration of additive (LMD) and subtractive (milling) processes to create thin-walled prototypes with complex external components that are not possible to obtain using separate AM (LMD) and SM CNC processes. Other applications include repairing damaged components, creating gradient functional materials, producing long overhang components without additional supports, embedding sensors and making special internal channels. Among others, real-time NI RT control, a fibre-optic IR sensor for temperature measurement in the powder melting zone, a laser displacement sensor for head positioning and a vision system were used [13].

The additive machining modules, which are installed on various multi-axis CNC machine tools use several layer deposition techniques such as: FDM, SLM, EBM, DED (*laser cladding*), or LBDW (*laser beam deposition welding*). The general principle is to employ a laser metal deposition (LMD) process. Different nomenclature of laser techniques can be used by manufacturers of hybrid machine tools, e.g. LENS, DMD, DED, PBF (*powder bed fusion*) or POM (*precision optical manufacturing*) [15].

An important design problem which appears in embedding a DMD (*directed metal deposition*) system into a CNC machine tool was the use of standard tool holders, e.g. an HSK 63 taper for mounting the deposition head in the machine spindle (Fig. 12.4b). However, an additional distributor is required to supply power to the head when it is mounted in the spindle and to cut off power when the LMD head is replaced by a cutter. Due to the need for many machining, measuring and control functions, all

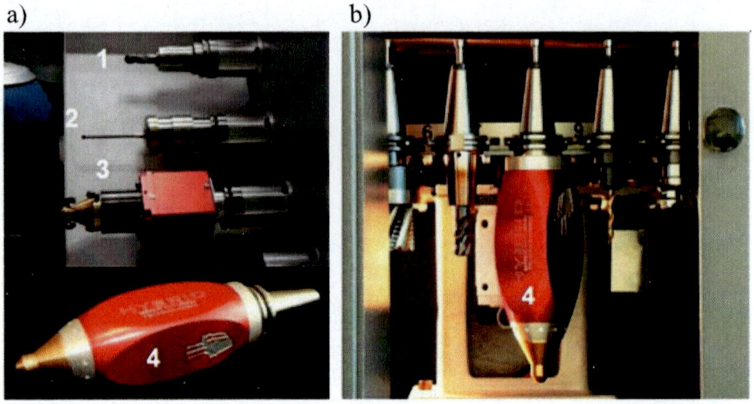

Fig. 12.9 Examples of tool magazine with cladding deposition head, ball-end milling cutter and touch probe (below the cladding head for powder deposition) (a) and vertical tool magazine on HSC hybrid turn-milling centre (b) [3, 14] (www.optomec.com). Symbols: 1—end mill cutter, 2—touch probe, 3—laser cladding head, 4—changeable laser cladding head

tools co-exist in the tool magazine with the tool changer, as for example shown in Fig. 12.9.

A versatile hybrid machine tool for hybrid forming, coating and repair uses the blown-powder deposition welding technique. This technique is more efficient than laser sintering and, for example, the production of a part of mass approx. 3.5 kg is 20 times faster while finishing milling is only used when justified. The machine tool is based on a 5-axis machining centre for precision milling with a 5-axis simultaneous machining capability. The AM unit is equipped with a 2.5 kW diode laser.

Examples of the technological capabilities and practical applications of this type of hybrid machine tools are shown in Fig. 12.10. The technological process of gradual part shaping is similar to the A/SM type process shown in Fig. 12.3. The cladding of cylindrical and conical parts is followed by CNC machining and finishing, e.g. polishing. It is possible to additively deposit bosses or helical coils and then finish mill the entire profile (Fig. 12.10a). The functionality of the hybrid machine tool is extended to perform turning, which enables the machining of rotational-symmetric components with subsequent milling of flashes (Fig. 12.10b). The use of a rotary table is a solution used in cladding or repairing compressor rotor blades for example as shown in Fig. 12.10c.

Figure 12.11a shows a vertical milling centre equipped with the LENS (*Laser Engineered Net Shaping*) incremental machining module, which has numerous applications in milling machines, lathes and robots as complementary to subtractive machining process. An example of a complex thin-walled part produced in a single setup is shown in Fig. 12.11b.

The hybrid machine tool which uses laser sintering of metal powder (3-D SLS) with an ytterbium fibre laser in the first stage and high-speed milling in the second stage is shown in Fig. 12.12 [4, 18]. What should be noted is that milling is carried

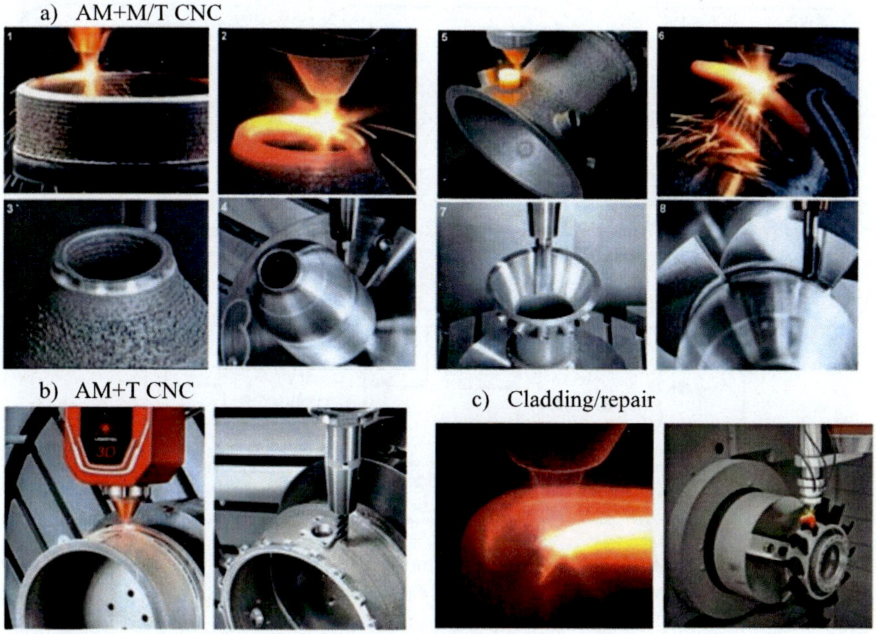

Fig. 12.10 Examples of laser cladding and milling (M) processes (**a**) or turning (T) CNC (**b**) and repair process of blades (**c**) [11, 16, 17] (www.dmgmori.com)

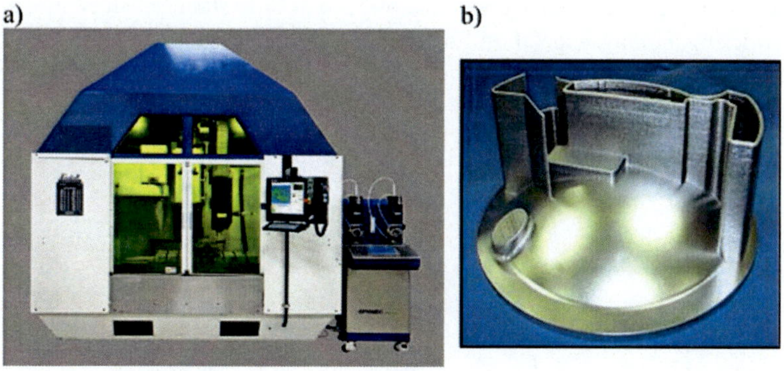

Fig. 12.11 Vertical milling centre equipped with LENS AM module (**a**) and example of a casing produced by a hybrid machining process (**b**) [14] (www.optomec.com)

out each time after 10 layers have been applied. The machine is equipped with a two-blade system for compacting the melted powder—the leading blade clears the path for the new layer, while the trailing blade smooths and compacts the freshly applied layer. The applied hybrid process allows significant cost savings and basically eliminates EDM from the mould and die manufacturing technology.

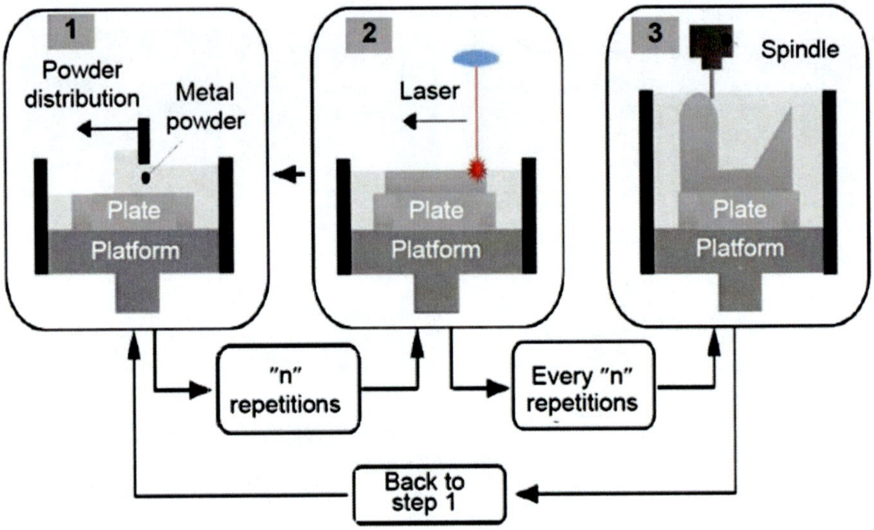

Fig. 12.12 Working scheme of hybrid vertical milling centre equipped with a laser sintering module [4, 18]. Symbols: 1—powder distribution; 2—laser processing; 3—milling

The technological capabilities of the machine are illustrated in Fig. 12.13a which is an injection mould made completely with a cooling channels system. For comparison, the injection mould shown in Fig. 12.13b was made from 18Ni (C300) dispersion-hardened steel powder with a grain size of 35 μm on a stainless-steel substrate by SLM sintering and high-speed milling (HSM).

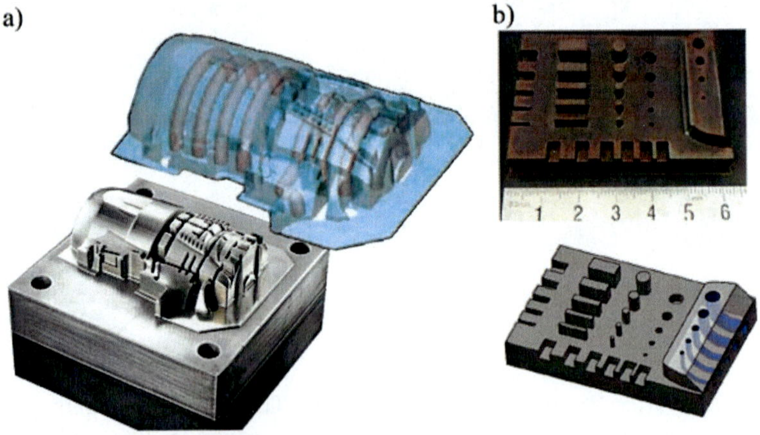

Fig. 12.13 Injection mould with internal cooling channels produced by A/SM hybrid process (**a**) and injection mould produced by SLM + CNC milling hybrid process (**b**) [3, 6, 8, 19]

A manufacturing platform, based on Laser Deposition Technology (LDT) which integrates laser deposition and 5-axis machining, enables hybrid fabrication, powder coating deposition and repair. It should be noted that fabrication can involve the creation of parts from a wide range of materials, e.g. copper/brass protrusions deposited to the circumference of an Inconel 718 alloy housing. The list of potential materials includes stainless steels, nickel-based alloys (Inconel 625, Inconel 718), bronze and brass alloys, chrome-cobalt-molybdenum alloys, stellite and weldable tool steels [6, 8, 21] .

The configurations of the two most commonly used hybrid manufacturing platforms in industry for the repair of worn/damaged parts are shown in Fig. 12.14. The solution in Fig. 12.14a is based on a turning-milling centre. The workpiece is fixed in the axis of the spindle, which allows for different orientation in relation to the tools during its rotation (the tool set is shown in Fig. 12.4a); additionally, special tools for polishing can be used [20]. An example of the implementation of hybrid machining built according to the concept in Fig. 12.15a is shown in Fig. 12.19. Figure 12.14b shows the solution of a hybrid manufacturing platform based on a 5-axis machining centre with an additional B rotational axis executed by a rotary-tilting table.

In the hybrid manufacturing platforms, employing the movements of AM and SM heads, are usually constrained to three controlled axes (X, Y and Z) and in order to incorporate more degrees of freedom (DOF) to the working heads, a 6-DOF robotic industrial arm (Fig. 12.15a) and changeable heads (Fig. 12.15b) is one of the possible proposals [22]. Such configuration of a multi-axis robotic platform (station) allows compound fabrication by integrating additive, subtractive and transformative processes. Figure 12.15c shows a concept of an integrated hybrid manufacturing platform equipped additionally with a heated bed for AM process (6-DOF printing) and clamping fixture for SM process. Several advantages of a 6-axis robotized hybrid manufacturing platform include reducing time and waste, improving the surface quality by eliminating the staircase error (see Fig. 12.1), manufacturing free-form surfaces, eliminating supporting structures and avoiding collision during manufacturing.

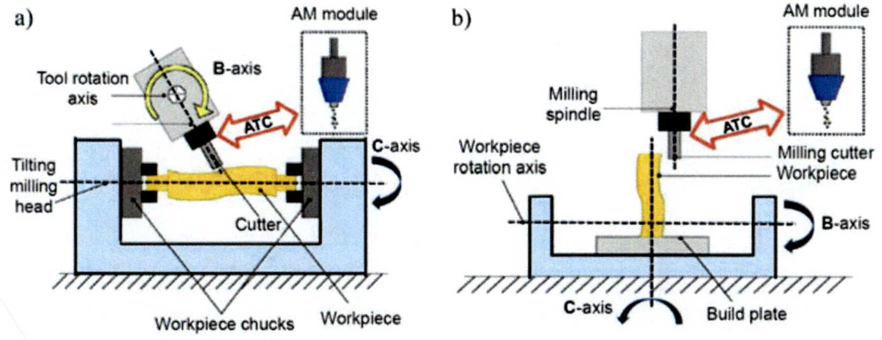

Fig. 12.14 Configurations of hybrid manufacturing platforms: on the base of turn-milling centre (**a**) and on the base of 5-axis machining centre (**b**), ATC- automatic tool changer [9]

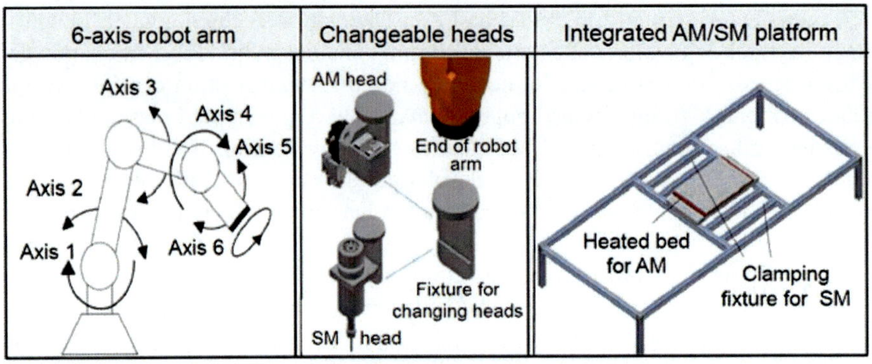

Fig. 12.15 Configuration of a robotized hybrid manufacturing platform: **a** 6-axis manipulator, **b** a set of changeable heads, **c** view of the hybrid platform [22]

12.2.2 Programming of Multi-axis Hybrid Machining Processes

The general concept of programming of hybrid AM + SM CNC machining is illustrated in Fig. 12.16. At the same time, on the basis of the created 3D CAD model, the part is printed according to the sent data in STL format and a CNC control program is created in the CAM system used. In the next stage, the manufactured part is finish-machined on a CNC machine tool. An integrated programming system based on the NX CAM package can be used to prepare the complete hybrid machining using a digital twin of the workpiece, fixtures and entire machine. The system calculates laser paths for additive manufacturing and cutting tool paths for subtractive machining operations.

Figure 12.17 shows examples of simulations of cladding (single layer deposition) and finish milling of the external profile of a part using a ball- end milling cutter. A

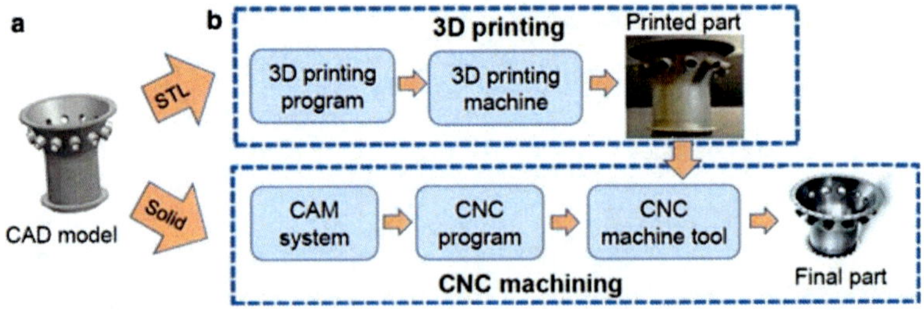

Fig. 12.16 General concept of integration **a** and complete programming **b** of additive and subtractive (AM + CNC) machining process according to Siemens [3] (www.siemens.com/plm)

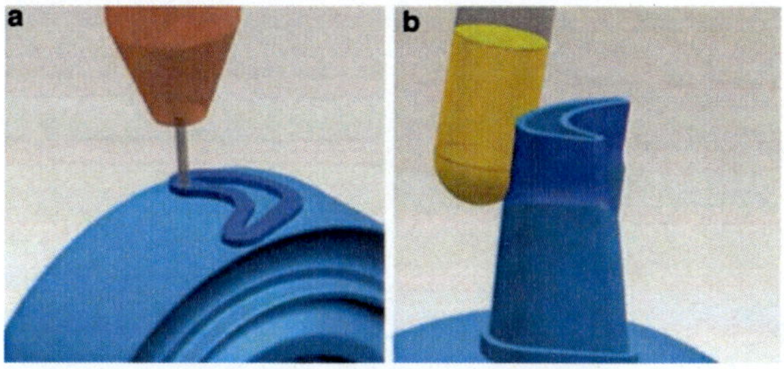

Fig. 12.17 Simulation *in-process* of additive shaping of part element (**a**) and verification of milling process (**b**) [3, 4] (www.siemens.com/nx)

special module available in the NX CAD/CAM package called *NX Hybrid Additive Manufacturing*, which can be easily configurated with a popular Siemens 840D CNC control system was used for this purpose [3].

12.3 Repair and Renovation Technologies

An important area of application for AM/SM CNC processes is the repair and restoration of functional properties to faulty pieces (e.g. undersized forgings) of large dimensions made from expensive materials. Special repair technology (LRT) is dedicated to this purpose.

A separate area of research and application of hybrid machining is the *remanufacturing of* expensive parts, mainly from the aerospace industry with large-scale use of heat-resistant and refractory materials (HRSA alloys), e.g. titanium alloy Ti6-4 and Inconel 718. In the initial stage of development, the focus was on the repair of turbine blades made of titanium Ti-6Al-4V with damaged edges and wear marks. Currently, laser repair technology (LRT), which includes several similar layer deposition techniques, such as: direct metal deposition (DMD), laser additive manufacturing (LAM) and laser metal deposition (LMD), is dynamically developed and finds numerous practical applications. In a nutshell, the method consists in removing material in the vicinity of the damage zone, applying new material to fill the defect and finishing by machining and polishing to restore the original geometry and functional properties of the surface.

Today, manufacturers offer specialised multi-axis hybrid machine tools integrating additive, direct deposition and subtractive machining processes, which are equipped with 3D laser scanners and inspection devices using 3D measuring probes. The total hybrid process thus includes the following activities: high-speed milling (HSM), 3D surface scanning, model creation and division of material loss layers, 3D laser cladding, 3D inspection, deburring/polishing and laser marking [20, 23, 24]. A detailed description of the repair cycle proposed by the RECLAIM project [12, 23] is shown in Fig. 12.18. It was proven that a hybrid machine tool with sufficient equip-

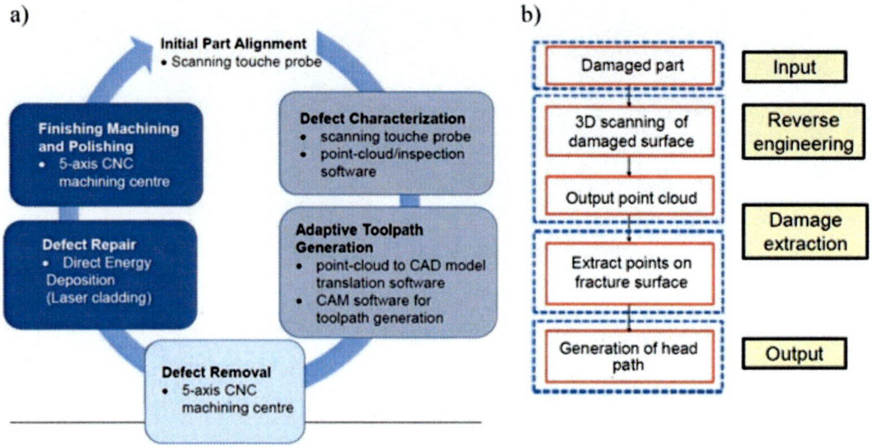

Fig. 12.18 Concept of hybrid repair technology on a multi-tasking machine tool (turn-milling centre) (**a**) and strategy for generation of deposition head path (**b**) [16, 17, 23]

ment is the most flexible system for renovating used parts because it consumes only a fraction of the energy, time and cost required to produce a new part.

Some examples of other large and expensive worn or defected parts selected for repairing, e.g. a turbine blade, an Inconel 718 alloy compressor housing, a planetary gear shaft and connector with a deposited layer 420 SS stainless steel and a defective Ti6-4 titanium alloy forging are shown in Figs. 12.19 and 12.20 respecively .

Fig. 12.19 Layer deposition on the blade edge (**a**) and milling of turbine blade after deposition (**b**) [20] (www.catapult.org.uk)

Fig. 12.20 Examples of repair processes of different worn/damaged parts using LDT (*Laser Deposition Technology*) technique [16, 17] (www.dmgmori.com). 1—worn part, 2—part after layer deposition, 3—part after machining of the weld

12.4 Development Trends in Hybrid Additive-Subtractive Processes

In recent years, there has been a significant use of hybrid additive and subtractive machining in various application areas, which is due to at least several reasons highlighted previously [3, 19, 23], i.e. producing parts with the desired accuracy and surface quality, adding complex fragments (e.g. stubs in Fig. 12.10b), which reduces costs, performing repairs in a single setup, producing parts from multiple materials by 3D printing.

The additive machining modules, which are installed on various multi-axis CNC machine tools use different deposition techniques such as: FDM, SLM, EBM, DED (*laser cladding*), or LBDW (*laser beam deposition welding*). The general principle is to use a laser metal deposition (LMD) process.

Depending on the performance requirements, parts are produced in manufacturing systems enabling different additive processes [15]. In the aerospace industry, a leading concept is the integration of the laser metal deposition (LMD) process into CNC milling centres, the use of light metal powders and the innovative manufacture of thin-walled structures. For the production of geared shafts, the laser in the LMD process is used for tooth deposition (milling is used for finishing) and additionally for surface hardening. Because the shaft is hollow, re-fixing operations are eliminated. In the case of the manufacturing of less loaded large-size forming tools, a combination of Gas Metal Arc Welding and milling on a CNC milling centre is more suitable. On the other hand, when heavily loaded tools, such as hot forming dies, are manufactured, the integration of platform powder deposition (LMPD-Laser Metal Powder Deposition) and milling on a CNC milling centre seems to be suitable [23].

An important and growing area of application for AM/SM CNC processes is the repair and restoration of functional properties to faulty pieces (e.g. undersized forgings) of large dimensions made from expensive materials. Special repair technology (LRT) is used for this purpose. An important issue is the integration of additive and subtractive processes to optimise hybrid machining because repairs usually involve expensive parts. As shown in Fig. 12.21, two approaches are possible. The first is when a new part is manufactured (in which case the layer deposition is followed by finishing machining by cutting), while the second is when it is necessary to repair damage or apply a coating (in which case alternative additive and subtractive operations are possible) [4].

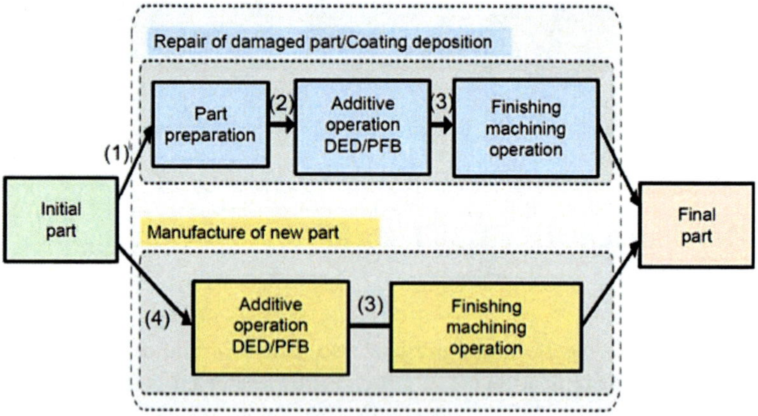

Fig. 12.21 Possible interaction within a hybrid machine for AM/SM hybrid processes [4]

Today, manufacturers offer specialised multi-axis hybrid machine tools integrating additive, direct deposition and subtractive machining processes, which are equipped with 3D laser scanners and inspection devices (3D measuring probes). The integration also includes CAD/CAM and CAI (*Computer Aided Inspection*) packages. As presented earlier in Sect. 1.6, hybrid machine tools should be reconfigurable and each process should be performed in a close loop system. The adaptability of the deposition module to the variety of materials, process conditions and geometries of the parts produced is necessary [23]. This, in turn, requires adequate metrological capabilities of measurement and monitoring systems to perform part inspection prior to process change. The future vision concerns intelligent machines on which not only different manufacturing processes but also assembly processes will be carried out. As a result, manufacturing, repair and renovation processes will be carried out on a single hybrid manufacturing station.

References

1. Grzesik W (2018) Hybrid additive and subtractive manufacturing processes and systems: a review. J Mach Eng 18(4):5–24
2. Grzesik W (2018) Advanced machining processes of metallic materials. Elsevier, Amsterdam
3. Jones JB (2014) The synergies of hybridizing CNC and additive manufacturing. Hybrid Manufacturing Technologies Ltd
4. Cortina M, Arrizubieta J, Ruiz JE, Ukar E, Lamikiz A (2018) Latest developments in Industrial hybrid machine tools that combine additive and subtractive operations-review. Mater 11:2583
5. Manogharan G, Wysk R, Harrysson O, Aman R (2015) AIMS- a metal additive hybrid manufacturing system: system architecture and attributes. Proc Manuf 1:273–286
6. Du W, Bai Q, Zhang B (2016) A novel method for additive/subtractive hybrid manufacturing of metallic parts. Proc Manuf 5:1018–1030
7. Schmidt M, Merklein M, Bourell D, Dimitrov D, Hausotte T, Wegener K, Overmeyer L, Vollerstsen LGN (2017) Laser based additive manufacturing in industry and academia. Ann CIRP Manuf Technol 66(2):561–583
8. Yamazaki T (2016) Development of a hybrid multi-tasking machine tool: integration of additive technology with CNC machining. Proc CIRP 42:81–86
9. Lee W, Wei Ch, Chung S (2014) Development of a rapid prototyping system using low-cost fused deposition modeling and five-axis machining. J Mater Proc Technol 214:2366–2374
10. Mazak's VC-500 AM hybrid speeds multitasking (2016) Additive manufacturing. www.mazak.com
11. Additive manufacturing in milling quality (2017) www.dmgmori.com
12. Reclaim Project – Remanufacturing the Future (2012) www.catapult.org.uk
13. Nagel JS, Liou FW (2012) Hybrid manufacturing system. Design and development. In: Aziz FA (ed) Manufacturing systems (Chap. 11). InTech Europe Rijeka
14. Sottak T (2018) Hybrid manufacturing. 3D printed metals—core applications. www.optomec.com
15. Merklein M, Junker D, Schaub A, Neubauer F (2016) Hybrid additive manufacturing technologies—an analysis regarding potential and applications. Phys Proc 83:549–559
16. Laser Deposition Technology (LDT) (2018) Examples. www.rpm-innovations.com
17. Grzesik W (2018) Hybrid manufacturing of metallic parts integrated additive and subtractive processes. Mechanik 91(7):468–475
18. Technology Spotlight (2018) Hybrid additive/subtractive machining centre. Matsuura Machinery. www.matsuurausa.com

19. Liu J, Wang X, Wang Y (2017) A complete study on satellite thruster structure (STS) manufactured by a hybrid manufacturing (HM) process with integration of additive and subtractive manufacture. Int J Adv Manuf Technol 92:4367–4377
20. Hybrid Manufacturing (2018) The world's first hybrid turbine blade & turbo fan remanufacturing machine. www.hamuel.com
21. Przestacki D, Chwalczuk T, Wojciechowski S (2017) The study on minimum uncut chip thickness and cutting forces during laser-assisted turning of WC/NiCr clad layers. Int J Adv Manuf Technol 91:3887–3898
22. Li L, Haghigi A, Yang Y (2018) A novel 6-axis hybrid additive-subtractive manufacturing process: design and case studies. J Manuf Proc 33:150–160
23. Flynn JM, Shokrani A, Newan ST, Dhokia V (2016) Hybrid additive and subtractive machine tools–research and industrial developments. Int J Mach Tools Manuf 101:79–101
24. Zhang X, Li W, Chen X, Cui W, Liou F (2018) Evaluation of component repair using direct metal deposition from scanned data. Int J Adv Manuf Technol 95:3335–3348

Chapter 13
Economics and Optimization Strategies of Hybrid Processes

13.1 Optimization Criteria and Models

Economic aspects are obviously important in the design of technological chain and planning a machining process taking into account costs involved. Among the total costs such components as the machining cost, tool cost, tool change and its setup costs, and handling costs are considered in the cost analysis. The optimization of machining parameters usually concerning single-pass machining operations can be performed based on two different points of view [1]:

- Minimum cost of the machining operation.
- Minimum time or equivalently maximum productivity of the machining operation (the reciprocal of the unit time is equal to the productivity, i.e. number of parts per time unit).

In machining practice some important optimization criteria are selected and not all machining parameters are optimized together. Mostly the function $T = f(v_c)$ is taken into consideration because tool life (T) influences predominantly the cutting speed (v_c) and, as a result, minimum operation cost or maximum productivity. Sometimes, such additional criteria as surface roughness or surface integrity are involved into optimization procedure.

In the case of hybrid machining processes additional variables, which determine the positive effects of hybridization, obviously in relation to productivity (machining removal rate) and surface finish are optimized [2–4]. They are as follows:

1. In the Assisted Hybrid Processes:

   - CF pressure in high pressure cutting (HPC),
   - heating temperature or heat source power in laser and plasma assisted processes,
   - frequency and amplitude of vibrations in vibration-assisted machining (VAM/USAM) operations,
   - type, temperature and supplying method of cryogenic medium,

W. Grzesik and A. Ruszaj, *Hybrid Manufacturing Processes*, Springer Series in Advanced Manufacturing, https://doi.org/10.1007/978-3-030-77107-2_13

- current and voltage, the width of discharge gap and the rotation speed of the wheel in abrasive assisted EDM (AEDG), electrolyte flow rate, voltage, current density and the rotation speed of the wheel in abrasive assisted ECM (AECG).

2. In the Mixed Hybrid Processes:

   - distribution of current and voltage impulses in electrochemical electrodischarge machining (ECDM),

3. In the Hybrid Processes with Controlled Mechanisms:

   - the depth of cut, feed rate, feed direction and tool geometry in turn-milling operation,
   - in-feed and grinding speed in grind-hardening operations,
   - normal load, burnishing speed and feed, pressure of cryogenic medium (LCO_2) and diameter of the burnishing element in deep rolling.

13.2 Optimization Criteria and Algorithms for the Selection of Machining Conditions

In conventional machining cost and time of individual machining operations are assessed based on the total minimum cost per piece (C_u) and the total minimum time per piece (t_u) as a function of the cutting speed (v_c). Consequently, as shown in Fig. 13.1, corresponding values (*mc*) or equivalent values of the productivity rate

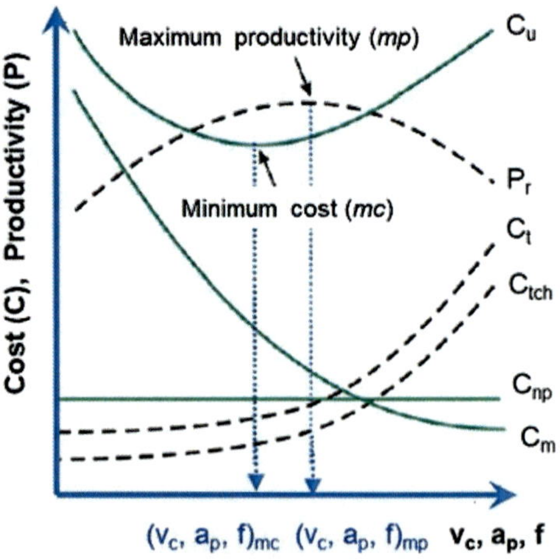

Fig. 13.1 Determination of optimal machining parameters for criteria of minimum cost and maximum productivity [1, 5]. Symbols: C_u—total cost per piece, P_r—production efficiency (production rate), C_t—tool cost, C_{tch}—tool change cost, C_m—machining cost, C_{np}—non-productive costs (constant)

(P_r) (*mp*) as functions of the cutting speed (v_c) are determined. These function can be extended to the optimal depth of cut and/or the optimal feed rate [1]. The optimal values of *mc* and *mp* are determined as the extremal points for the functions $C_{uopt} = f(v_c,a_p,f)$ and $P_{ropt} = f(v_c,a_p,f)$ for selected machining conditions. The structure of costs and times consider the components related to the machine tool, the tool and tool manipulation. The cost values include the machining cost (C_m), the tool cost (C_t), the tool change cost (C_{tch}) and non-productive cost (C_{np}), i.e. the total cost per piece is equal to $C_u = C_m + C_t + C_{tch} + C_{np}$. Consequently, the productivity rate is expressed as $P_r = f(t_m + t_{tch} + t_{np})$.

The optimization procedure employed in the strategy of sustainable machining/manufacturing includes not only cost and time but also energy consumption and utilization of material resources [1]. This implies that in innovative hybrid machining processes the multi-criterion optimization should be performed. In the case of hybrid machining these criteria can be related to costs and energy efficiency criterion (obviously energy consumption) [6]. It is interesting and practically important to consider different cooling strategies when media-assisted hybrid processes are implemented.

Figure 13.2a presents a comparative assessment of energy consumption and total cost components as a function of the MRR when drilling in iron of EN-GJL 250 grade using optionally dry machining, emulsion cooling, MQL technique and cryogenic machining with CO_2 snow. It is depicted in Fig. 13.2a that higher MMR generates higher tool costs due to shorter tool life. In this aspect, emulsion cooling or cryogenic machining seems to be the most efficient but in the second case the energy consumption per hole is distinctly lower (Fig. 13.2b). However, the cryogenic machining is the first choice because the increase in MRR corresponds to a lower energy consumption.

As can be seen in Fig. 13.2a, the minimum of total costs can be obtained for different cooling strategies but the maximum productivity corresponding this optimization criterion appears also for cryogenic cooling using CO_2 snow (it is also valid for lower process efficiency characteristic for MQL system). On the other hand, wet machining using emulsion and MQL aerosol can be recommended for medium and

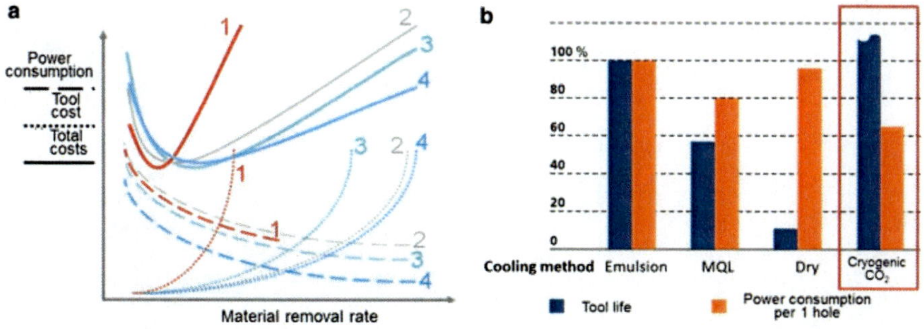

Fig. 13.2 Comparison of power consumption and costs when testing different cooling strategies: 1—dry, 2—wet (emulsion), 3—MQL, 4—cryogenic (**a**) and tool life (**b**) in drilling EN-GJL 250 grey iron [5] (www.iwu.fraunhofer.de)

dry machining for lower productivity. The analysis of cooling strategies done indicates that cryogenic machining is more sustainable because the profits cover not only lower machining cost and shorter machining time but also lower environmental impact and energy consumption.

In the case of machining extremely difficult-to-machine materials such as aerospace heat resistant alloys, the material removal rate (MRR) can only be, in practice, improved by using HPC operations (Sect. 7.3). For instance, as shown in Fig. 13.3, the increase of MRR by 100% in the machining of Inconel 718 alloy was achieved for coolant pressure of 150 bar. Additional advantages observed are better chip control and longer tool life due to more intensive cooling effect of the tool. As a result, the process behaviour becomes more stable and reliable. Similar technological effects are obtained in the milling operation with CF supply through internal channels in the cutter (Fig. 13.4) instead of traditional technological chain consisting

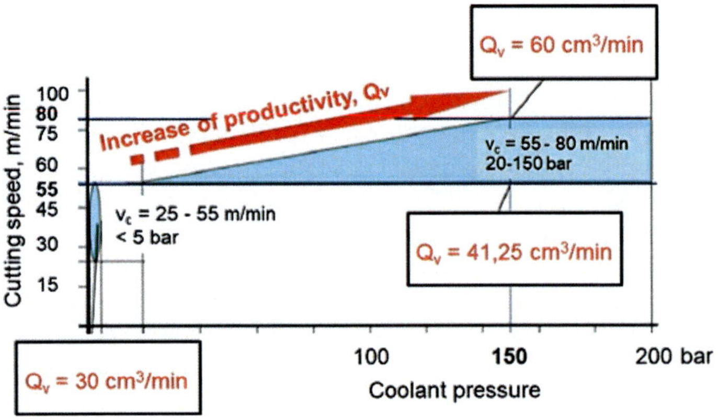

Fig. 13.3 Increase of material removal rate in machining Inconel 718 with high pressure coolant (HPC) [7]

Fig. 13.4 Comparison of costs and machining time for three variants of pressure die machining [5]

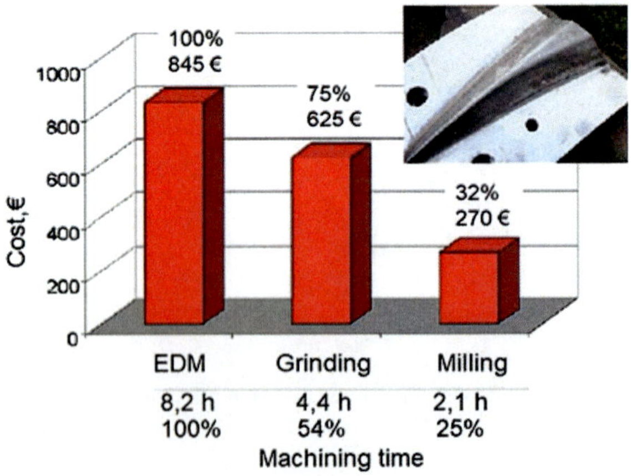

of roughing EDM and finishing grinding operations.

Figure 13.4 presents comparison of time, cost and resource effectiveness when machining pressure die using EDM, grinding and hard milling operations. It is evident from the diagram shown in Fig. 13.4 that the application of finish hard milling assisted by high pressure coolant delivered through the inner channels in the milling cutter (HPJC) allows to save of 75% machining time and reduce costs by 68%.

Figure 13.5 presents the three fundamental pillars of sustainable concept of manufacturing/machining activities including economy, environmental and society [1, 8]. It is essential to distinguish the four life cycle stages, i.e. pre-manufacturing, manufacturing, use (service) and post-use. However, the development of sustainability is based on the appropriate models and factors taking into account a holistic and objective assessment algorithm. In particular, the quantitative metrics of process sustainability allows to select the most convenient process behaviour among different cooling strategies, i.e. dry, wet (flooding), MQL and cryogenic using LN2 jet. The selection of appropriate cooling and lubrication strategies is a decisive concern in machining processes. In order to optimise the sustainability performance, such indicators as the Process Sustainability Index (*ProcSI*), Product Sustainability Index (*ProdSI*), Total Weighted Sustainability Index (*TWSI*) and Technique for Order Preference Based on Similarity to Ideal Solution (*TOPSIS*) have been introduced [8]. The TOPSIS algorithm along with the entropy weight method (*EWM*) allows to generate the value R_i ranging from 0 to 1 which is called the Relative Sustainability of Process (*RSP*). A higher value indicates that the alternative is closer to the positive ideal.

In the diagram presented in Fig. 13.6 the results of 20 experiments are compared using the RSP factor as the sustainability measure. In this comparison such variable factors as the tool-chip temperature, specific cutting energy, surface roughness

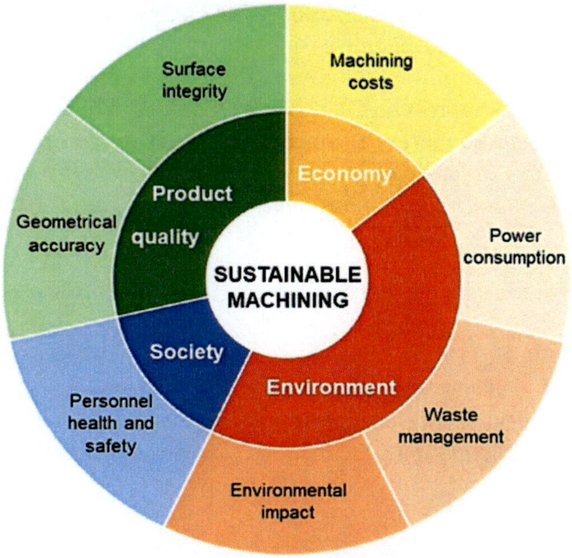

Fig. 13.5 Circle diagram of sustainable machining process [8]

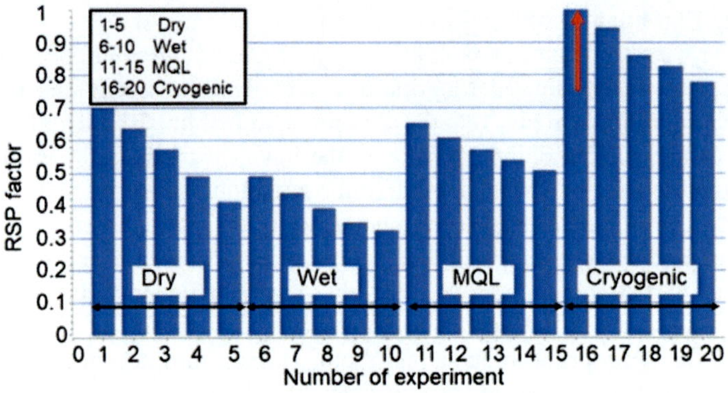

Fig. 13.6 Comparison of RSP factor for different cooling strategies [8]

and environmental impact due to electric energy consumption (expressed in kg-CO_2/mm^3) are considered. The distribution of the RSP factor confirms that the use of LN2 as the cryogenic medium (experiments 16–20) outperformed other cases in terms of tool wear and surface roughness. The experiment number 16 was carried out with the smallest depth of cut of 0.2 mm. The performance characteristic corresponding to these machining conditions are: average contact temperature $T = 24$ °C, tool flank wear $VB_B = 55$ μm, tool rake wear $KT = 62$ μm and surface roughness parameter $Ra = 1.3$ μm (Fig. 13.6).

Investigations of energy consumption in manufacturing processes and saving of natural resources are obviously extended into the problems of the influence of energy consumption on the process performance and the relationships between process conditions and possible environmental impact. Because manufacturing has a complex hierarchical structure, an energy consumption analysis is basically limited for individual processes including machining processes-conventional, unconventional, hybrid and additive. According to the authorities it is necessary to use an integral approach called *eco-efficiency of manufacturing process* [1, 9]. The term *eco-efficiency (Eeff),* which corresponds to the sustainable manufacturing strategy, means „*create more value with less environmental impact*". It was defined by *World Business Council of Sustainable Development* (WBCSD) as follows [1, 9]:

$$Eeff = \frac{product\ or\ sevice\ value}{environmental\ impact} \tag{13.1}$$

In Eq. (13.1) the numerator gives the number of produced/sold products or the value of net sale, and the denominator considers different influences on the environment such as energy, water and material consumption and also greenhouse gas (GHG) emission or reduction of ozone concentration. In the case of manufacturing processes the assessment of eco-efficiency can be based on *Life Cycle Cost/Life Cycle Assessment*- LCC/LCA.

13.3 Optimization Fundamentals

The rapid development of hybrid manufacturing processes and hybrid manufacturing systems causes that more and more planning operational systems are employed. It should be noticed that design of hybrid machining processes is a multi-disciplinary problem and its successful solving is based on the physics of constituent processes as well as the design of dedicated machine tools. A new challenge is the development of systematic and scientific approach to the grouping, description and explanation of rules for integrating individual machining/manufacturing processes. One of the possible design methodology is based on the TRIZ (*Theory of Inventive Problem Solving*) algorithm [10, 11]. The original acronym TRIZ was proposed in Russian language.

The flowchart for general design methodology of a hybrid machining process is illustrated in Fig. 13.7. It outlines the structure of design methodology and subsequent steps which are required for designing innovative hybrid manufacturing processes. In the first step the limitations of the existing machining processes are formulated and, then, the possibilities of another process capable of addressing this limitation are verified. Next, hybridization can be conceptualized based on the gathered knowledge on two selected processes and their technological capabilities. It is very important to decide if these processes interact with each other, if the energy sources are compatible,

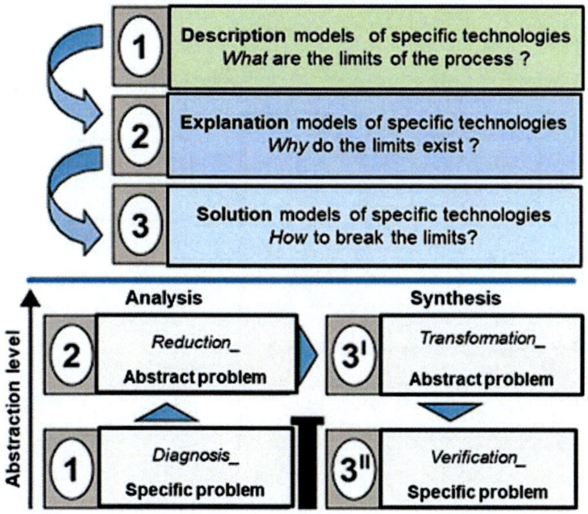

Fig. 13.7 Steps in design methodology of hybrid manufacturing processes based on similarities in the TRIZ algorithm [2, 10] (www.triz-journal.com)

and their influences are synchronized. In the subsequent step, process analogies which assist in the selection of a specific solution are provided. They are based on the grouping of solutions and mechanisms of utilized machining/manufacturing processes as well as individual conventional and hybrid processes. For instance, when local heating of the workpiece surface is required, the specific solution will be a laser beam. In practice, there is a need for individual process optimization in terms of the selection of the tool and machining conditions. This approach to process improvement is solved using conventional methodology, as shown in the upper part of flowchart in Fig. 13.8. However, it should be evident that any important changes in the applied machining technology cannot be achieved. As mentioned above, only an interdisciplinary approach can be effective, as depicted in the lower part of Fig. 13.8.

It is evident from Fig. 13.8 that after exhausting of the capabilities of a conventional optimization, the process behaviour can be improved only by applications of additional outer energy sources, i.e. the application of equivalent hybrid assisted

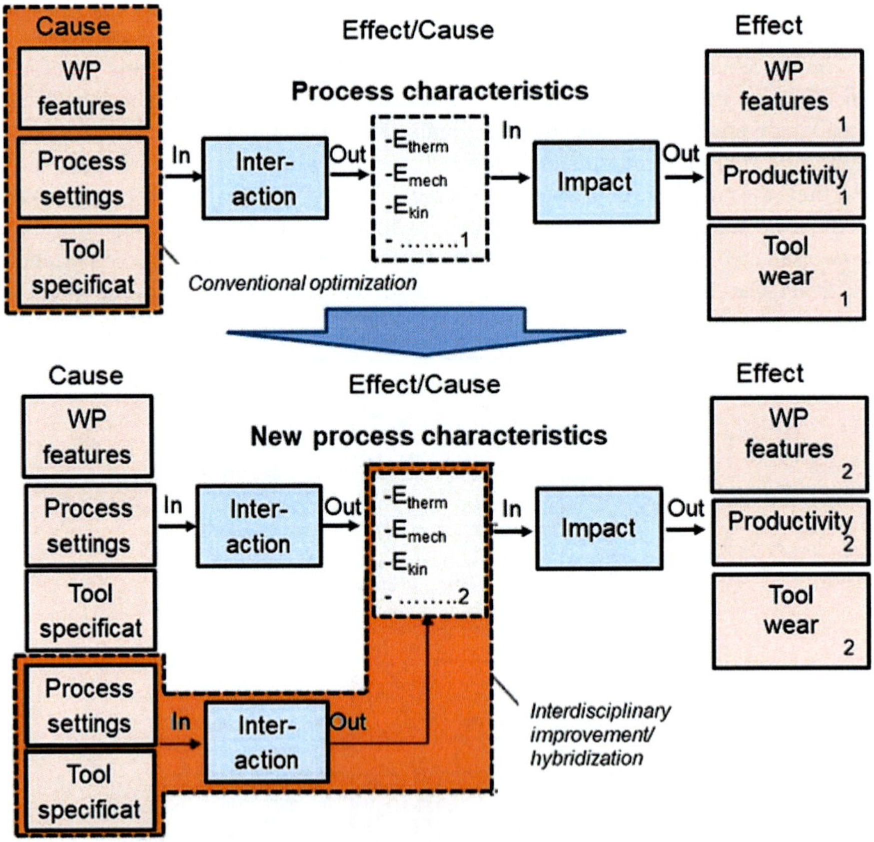

Fig. 13.8 Comparison of conventional optimization with interdisciplinary effect of process hybridization [1, 10, 11] (www.triz-journal.com)

processes. The procedure implemented is based on the close interaction of the mechanisms of new assisted processes in relation to the main (primary) conventional process. Another possibility is to employ the *effect-cause* type analysis related to all elements of the technological chain, perform the integration of unit/individual processes or use the cumulative effect of several processes in one hybrid process.

As presented in Sect. 12.4 manufacturing hybrid technologies integrating additive (AM) and subtractive (SM) processes, as well as repair and renovation technologies, are not yet satisfactorily developed and frequent operator's interventions for the generation and interpretation of features of constituent processes during process planning are necessary. In addition, less attention is devoted to critical decisions regarding available different versions of process plans, i.e. selection of most economically convenient process plan [12]. In the cost-driven method presented schematically in Fig. 13.9, the economical assessment of a hybrid process of an end-of-life part as a semi-finished part for further use (*remanufacturing*) is based on its end of life. The methodology presented in Fig. 13.9 includes 4 stages. In the first stage, features of AM and SM processes are recognized using a special feature recognition algorithm which explores geometrical information on the new part. It is noticed that an automatic feature recognition utilizes the *level-set-method*, which is successfully applied in the vision system for segmentation and recognition of digitized images. Based on the obtained information on process characteristics and process planning rules, in stage 2, partial process plans for the selected parts are generated. Some undetermined sequences of the plans are supplemented using a solution computed by means of a *mixed integer* programming model. The process plan for each scenario is then optimized locally. In stage 3, the selection of the most convenient cost-driven variant of the process and a comparison of costs for each process scenario are carried out.

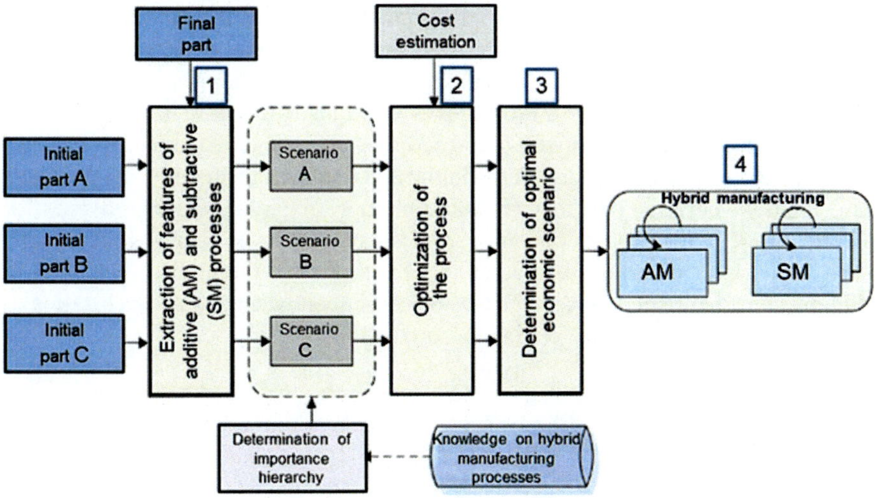

Fig. 13.9 Method of cost-driven planning method for additive-subtractive hybrid remanufacturing process [12]

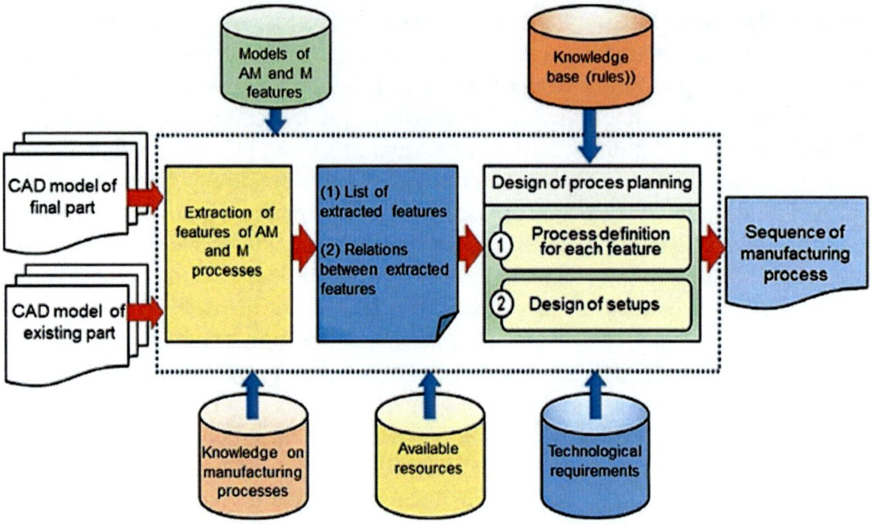

Fig. 13.10 Design methodology of a sequence of additive-subtractive hybrid process [13]

As a result, the selected hybrid machining (HM) process (stage 4) will be performed according the most effective process plan. In work [12] a case study of the hybrid remanufacturing process using the *Powder Bed Fusion* (PBF) technique in the AM part and 3-axis CNC machine tool in SM part is presented.

Figure 13.10 presents the procedure for the generation of additive-subtractive hybrid process which allows producing the final part from the delivered semi-finished part. The procedure is based on the technological requirements which can be directly available from the part drawn, the knowledge on manufacturing processes, i.e. AM process of metallic parts and CNC machining, and available setups i.e. AM machines, CNC machine tools, cutting tools and fixturing system. The design of sequences of manufacturing processes is done in two stages. In the first stage features/characteristics of subtractive and additive processes are searched from available data bases and CAD model for the initial and final part geometries. At this stage some relations between part features and limitations regarding the effectiveness of material usage in the process planning are available. In the second stage the generated features and relations are used as the input data for the process planning. In addition, some heuristic rules, which will be consequently used in hierarchization of the sequences of AM and SM operations, are defined.

References

1. Grzesik W (2017) Advanced machining processes of metallic materials. Elsevier, Amsterdam
2. Luo X, Qin Y (2018) Hybrid machining. Theory, methods, and case studies. AP Elsevier, London
3. El-Hofy H (2005) Advanced machining processes. Nontraditional and hybrid processes. McGraw-Hill, New York
4. Lauwers B, Klocke F, Klink A, Tekkaya AE, Neugebauer R, McIntosh D (2014) Hybrid processes in manufacturing. Ann CIRP Manuf Technol 63(2):561–583
5. Neugebauer R, Drossel W, Wertheim R, Hochmuth R, Dix M (2012) Resource and energy efficiency in machining using high-performance and hybrid processes. Proc CIRP 1:3–16
6. Grzesik W (2013) Energy consumption optimization in machining processes. In: Davim JP (ed) Metal cutting technologies. Progress and current trends (Chap. 2). De Gruyter, Oldenbourg, Berlin
7. M'Saoubi R, Axinte D, Leung Soo S, Nobel Ch, Attia H, Kappmeyer G, Engin S, Wei-Ming S (2015) High performance cutting of advanced aerospace alloys and composite materials. Ann CIRP Manuf Technol 64(2): 557–580
8. Bhat P, Agrawal Ch, Khanna N (2020) Development of a sustainability assessment algorithm and its validation using case studies on cryogenic machining. J Manuf Mater Proc 4(2):42
9. Li W, Winter M, Kara S, Hermann Ch (2012) Eco-efficiency of manufacturing processes: a grinding case. Ann CIRP Manuf Technol 61(1):59–62
10. Klocke F, Roderburg A, Zeppenfeld (2011) Design methodology for hybrid production processes. Proc Eng 9: 417–430
11. Roderburg A, Klocke F, Zeppenfield Ch (2009) Design methodology for hybrid production processes. TRIZ J. www.triz-journal.com/archives/2009/04/02/
12. Zheng Y, Liu J, Ahmad R (2020) Cost-driven process planning method for hybrid additive–subtractive remanufacturing. J Manuf System 55:248–263
13. Le V, Paris H, Mandil G (2017) Process planning for combined additive and subtractive manufacturing technologies in a remanufacturing context. J Manuf System 44:243–254

Chapter 14
Influence of Process Hybridization on Surface Integrity

14.1 Structural Models of Subsurface Layer

In material removal processes the material layer localized beneath the real surface, which is affected by different process alterations including physical (heat, forces, diffusion or foreign atoms), chemical (i.e. oxidation) and structural (metallurgical changes, phase transformations) influences is termed *subsurface layer or superficial layer* (SL) [1]. On the other hand, more general term *surface integrity* describes the topological (geometrical) features of the machined surfaces and their physical and chemical properties, and also the mechanical and metallurgical properties of the subsurface layer. From the metrological point of view the subsurface layer is defined as a material layer with altered mechanical/physical and sometimes chemical properties during the machining process in relation to the bulk/core material. In order to characterize the process influence on the properties of the SL the structural models of the SL with different number of specific zones (from 3 to 8 individual sublayers) are proposed [1–3].

Figure 14.1 presents the simplest three-zone model consisting of the external zone with different contaminations and absorbed gases, the middle zone with altered micro-structure and texture, and the internal zone, containing less deformed material grains, which differs visibly from the core material. The most important zone, from the technological point of view, is the middle zone within which material is highly deformed and thermally affected, and, as a result, its properties differ distinctly from the bulk material. The thickness of the middle layer ranges from 0.5 to 500 μm depending on the process conditions. It should be noted, that mechanical properties of the middle layer have a predominant influence on the functional properties of the machined parts [1, 4]. The state and attributes of a real manufactured surface that influences functional performance, such as fatigue strength, corrosion resistance and service life, are globally characterized by surface integrity [1, 3, 5]. It should emphasized that process hybridization can be utilized more effectively in the generation of the demanded surface integrity of machined parts [6].

Figure 14.2 illustrates possible distributions of SL properties determined exper-

W. Grzesik and A. Ruszaj, *Hybrid Manufacturing Processes*, Springer Series in Advanced Manufacturing, https://doi.org/10.1007/978-3-030-77107-2_14

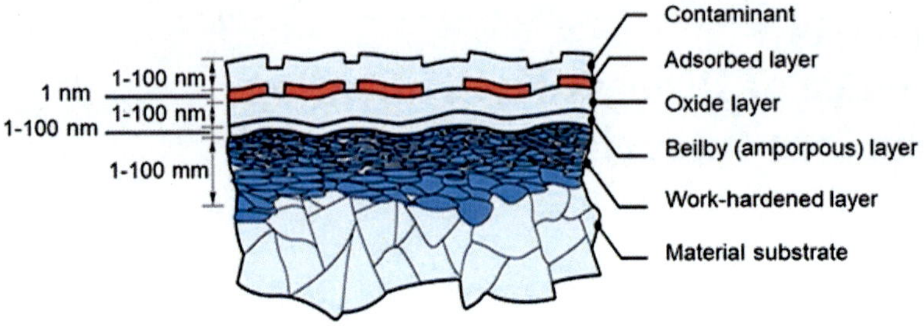

Fig. 14.1 Characteristic zones of three-zone structural model of the subsurface layer [1, 2]

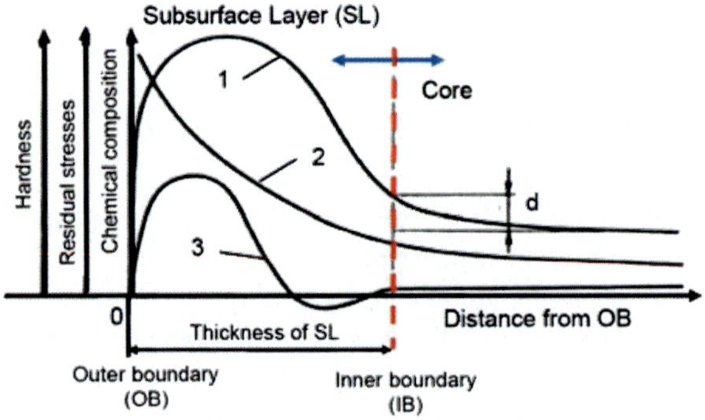

Fig. 14.2 Schematic distribution of SL properties within metrologically distinguished thickness. Symbols: 1—hardness, 2—chemical composition, 3—residual stress

imentally at a distance between the outer (OB) and inner (IB) boundaries. They concern the strain-hardening effect which influences material hardness, residual stresses which are induced due to both mechanical and thermal effects, and microstructural alterations which result from the thermal effect and associated phase transformations. It is not possible to determine sharply the localization of the inner boundary and a 20% deviation (d) is obviously assumed based on the experimental results [1].

The classical approach to the possible correlations between surface finish/surface integrity and the post-process surface functionality is presented in Fig. 14.3. In practice, two types of correlations are used. The first approach is to define the functional relations between machining conditions (as variable factors concerning machining parameters, colling/lubrications methods and cutting tool characteristics). In the next step, surface functionality can be predicted. The second approach, which is the most important from the surface functionality, is the correlation between process conditions and demanded functional properties. In a more general consideration, these functional relations can be related to the general structural model of the material removal process, i.e. to the machine-fixture-workpiece-tool (*mfwt*) technological

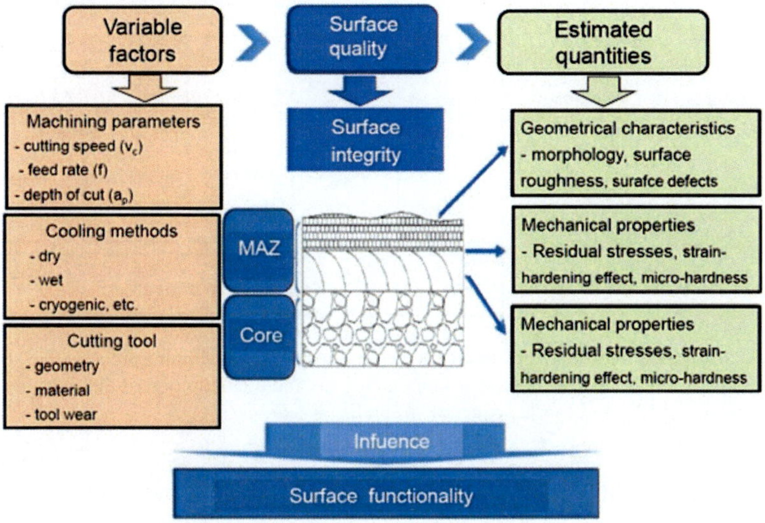

Fig. 14.3 Correlations between surface quality and machining conditions and properties of the subsurface layer (SL) [3, 7]. Symbol: MAZ/TAZ—mechanically/thermally affected zone

system. In such an approach, the decisive issue seems to be the selection of functional factors which allows correlating physical behaviour of the machining process and its influence on the SL state and corresponding functional properties.

It is documented in Sect. 13.5 that the final state of the machined surface and the sublayer depends on the individual mechanical, thermal and chemical loads or their combinations, which is crucial for the performance of hybrid machining processes. Problems related to different process interactions based on synergy effect which occur in hybrid machining processes are outlined in Chap. 13. Depending on the type of interaction between elements of the machining (*mfwt*) system the following ones are distinguished:

- outer mechanical and thermal loads exerted on the machined material,
- inner mechanical and thermal loads, i.e. deformation and temperature fields, generated inside the material during the process, i.e. during heating by laser or plasma.

As shown in Fig. 14.4, the outer and inner loads are represented by appropriate energy fluxes denoted by symbols E_{mech}, E_{therm} and E_{chem}. As a result of energy conversion and dissipation (for instance conversion of mechanical energy into heat generation and energy dissipation due to friction), the initial material state is modified and the total material alteration influences the final geometrical and physical state of the subsurface layer, i.e. surface integrity [1]. It can be noted that the time of process influence (action of specific loads) on the material is practically the same as for energy conversion and dissipation. Hence, it can be assumed, taking this principle into account, that from the physical point of view machining processes are performed

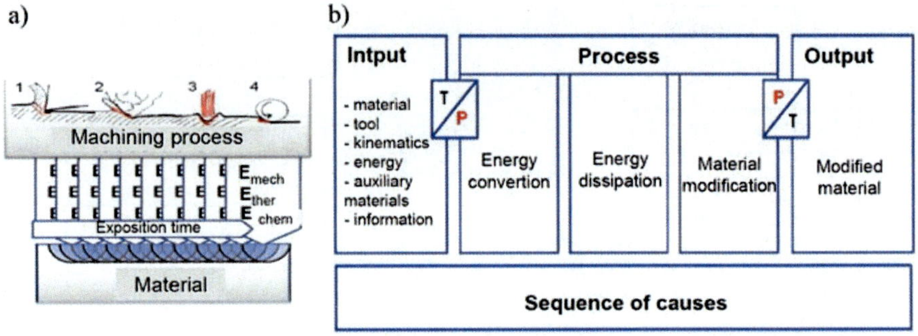

Fig. 14.4 Scheme of controlled functional interactions in different machining processes: (1) cutting, (2) abrasion, (3) LBM, PBM and IBM, (4) burnishing (**a**) and the definition of the structure of the physical process model and causal sequence (**b**) [4, 8]

as shown schematically in Fig. 14.4b. Finally, a sequence of causes, which results in controlled modifications of material properties, can be predicted.

The main information required for the formulation of the concept of a hybrid machining process is that each process is characterized by a unique combinations of material loads. Such processes as deep rolling, EDM and ECM represent one main type of load, i.e. mechanical, thermal and chemical respectively. On the other hand, conventional cutting and abrasive processes are controlled by a combination of mechanical and thermal loads.

14.2 Characteristics of Surface Roughness in Different Hybrid Machining Processes

It is well known that surface roughness or surface topography in a broader aspect can be generated by deterministic, random or mixed deterministic-random process interactions on the workpiece surface [1]. A higher participation of random interactions takes place in finishing cutting and abrasive operations. This fact is crucial for applications of both contact and non-contact measurements of surface roughness. In the engineering practice, both 2D and 3D surface roughness evaluation methods are used. In the case of 3D characterization, the surface texture or surface topography is characterized using appropriate S and V areal roughness parameters.

In typical cutting operations such as turning, milling and drilling the real machined surface consists of feed marks produced by geometrical-kinematical reproduction of the tool corner. This means that the shape of irregularities generated on the machined surface depends on the geometrical configuration of the tool corner as a geometrical factor and the feed rate as a kinematical factor. In precision and ultra-precision machining operations some non-geometrical micro-factors such as the cutting edge roughness/surficial defects or cutting vibrations should be additionally considered [1, 9].

In hybrid machining processes assisted by vibrations (Chap. 6), technological media (Chap. 7) or thermal energy (Chap. 9) the reduction of surface roughness can be achieved by controlling non-geometrical influences such as friction and elastic–plastic deformation (in particular spring-back effect resulting from elastic recovery of the hard material). In machining of hard and brittle ceramic materials it is necessary to soften thermally the subsurface layer and to machine it in the ductile state (Chap. 9).

Improvement of surface finish in the hybrid machining of MAM and LAM/IAM types is presented in Fig. 14.5a, b respectively. Supplying the coolant at the pressure of 110 MPa causes that the surface roughness parameter Ra is reduced even by 30% depending on the feed rate used. In the case of thermal assistance by laser or by both laser and induction (Fig. 14.5b) the improvement of surface finish depends on the heat source employed and the heating method used. Comparatively, the three different heating methods are tested, i.e. using separately laser and induction to the temperature of 900 °C, using combined heating by laser (LAM) and induction (IAM) to the temperature (M1-LAM to 900 °C and IAM to 700 °C, M2-LAM to 700 °C and IAM to 900 °C). It was revealed that the combined heating according to variant M1 for which the Ra parameter was reduced is more effective in comparison to simple laser source, by 25–35% [11]. It was explained by a more pronounced thermal softening of the subsurface layer resulting from the two-stepped heating.

Figure 14.6a presents the influence of laser pre-heating in a turning (LAT) operation of a Al-based composite (45vol% SiCp/2020Al) reinforced with SiC particles of 5–10 μm size, which is classified as a material with a very poor machinability rate [1], on the surface roughness in terms of feed rate. A fibre laser with the maximum power of 300 W (30 W in this study) was employed. It can be seen in Fig. 14.6 that the surface roughness obtained after LAT operation is substantially lower but the maximum reduction of Ra value was documented for the feed of 0.1 mm/rev and a small depth of cut of 0.05 mm and 0.1 mm (about 100% and 300% respectively).

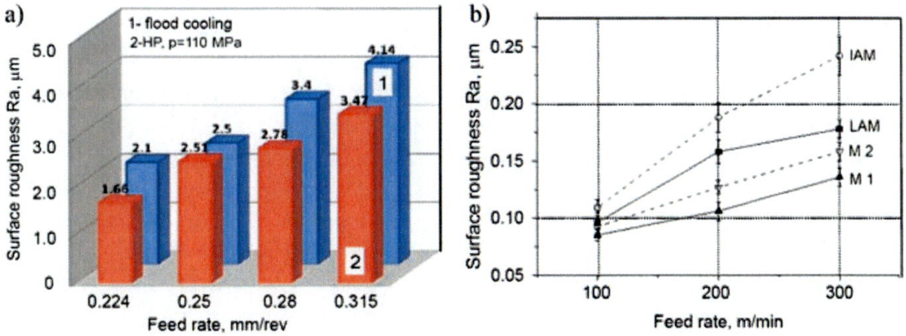

Fig. 14.5 Comparison of surface roughness parameter Ra after turning Ck45 (58HRC) steel using optionally flood and high-pressure jet assisted (HPJAM) with the coolant pressure of p = 110 bar: $v_c = 98.5$ m/min, $a_p = 2$ mm (**a**) [10] and after milling Inconel 718 with laser (LAM) and induction (IAM) material pre-heating [11]

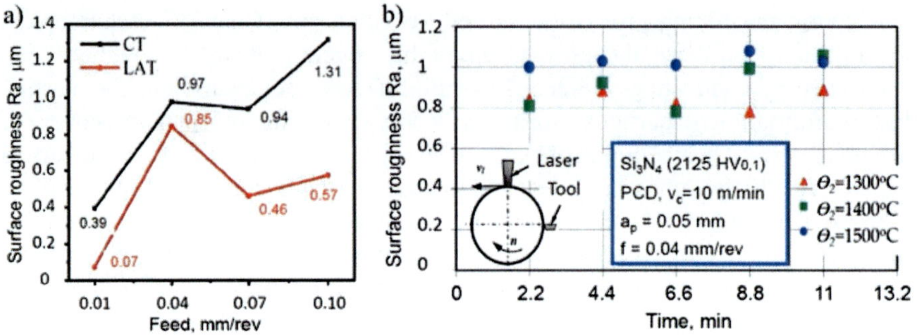

Fig. 14.6 Influence of laser pre-heating in the turning of a particle reinforced Al-based composite (45vol% SiCp/2020Al) on surface roughness in terms of feed rate ($v_c = 30$ m/min, $a_p = 0.05$mm) (**a**) [12] and nitride ceramics (Si_3N_4) with PCD tool (**b**) [13]

However, this trend was not observed for a higher depth of cut of 0.2 mm because the thermal softening effect weakens [12].

Figure 14.6b presents some results of measurements of the Ra roughness parameter when LAT process ($v_c = 10$ m/min, $f = 0.04$ mm/rev, $a_p = 0.05$ mm) was applied to a part made from nitride ceramics (Si_3N_4) using PCD tool and the molecular laser CO_2 of 2.6 kW power. It can be noted that the part pre-heating at $\Theta_2 = 1500$ °C results in the increase of Ra parameter by 20% and a lower temperature of 1300 °C is recommended to reduce surface roughness. Moreover, it was revealed that surface roughness practically does not change with the machining progress but depends on the pre-heating temperature.

In general, surface roughness produced by unconventional machining processes (EDM and ECM) depends on such physical/chemical processes as melting and vaporization, ablation or electrochemical dissolution [9]. In the case of EDM process the surface topography produced is of a random nature (P type) and consists of localized spherical craters generated by electric discharges. In particular, the material ratio is very low. Typically, surfaces with Ra = 0.8–3.2 µm (obtainable 0.25 µm) after finishing and Ra = 6.3–12.5 µm after roughing EDM operations are generated [9]. In ECM process material is removed “atom” by “atom” and, as a result, the surface roughness is distinctly lower in comparison to other unconventional processes. Typically, surfaces with Ra = 0.32–1.25 µm and rounded peaks of 90–220 µm radius are produced. The material ratio is comparable with the effect of face finish grinding [10].

14.3 Physical Properties of Subsurface Layer

14.3.1 Characteristics of Physical Properties of Subsurface Layer

The functional properties of the machined parts can be assessed based on characteristic mechanical quantities including [1]:

- Hardness (meso, micro and nano-hardness) as the measure of strain-hardening effect across the subsurface layer,
- Residual stresses generated according to mechanical and thermal loads exerted on the machined material.

It was mentioned many times that the hybridization of machining processes changes and enhances properties of the subsurface layer and allows adopting them better to the technological demands. Below, a short survey of experimental results concerning alterations of the machined surface and subsurface layer will be presented taking into account three basic hybridization methods, i.e. assisted by additional energy sources, combination of different processes and controlled mechanisms of main processes.

14.3.2 Residual Stresses in Subsurface Layer

The final state of residual stresses (RSs) induced to the subsurface layer is always the effect of superposition of mechanical and thermal stresses. When material strain-hardening resulting from mechanical loads is not sufficient and the negative plastic deformation occurs in the zone behind the tool, tensile residual stresses of mechanical character will be generated. Taking into account that the thermal mechanism always leads to the generation of tensile residual stresses, the cumulation of RSs of the same sign occurs. The same situation can be observed when phase transformation or predominant thermal influence are presented during the machining process. However, it can be reasoned, based on the previous investigations [10, 11], that some groups of machining/manufacturing processes have a similar potential regarding the generation of subsurface layer with demanded functional properties. Despite of distinctly different influences of the internal loads exerted on the material during the process, the final modification of the subsurface layer is similar in terms of the hardness and RS distribution profiles. This phenomenon based on the similarity of process outputs is shown in Fig. 14.7. It is evidently indicated in Fig. 14.7 that the distribution profiles of micro-hardness and residual stresses are quite comparable after induction and grind-hardening as well as after surface hardening by deep rolling and grind-strengthening.

Changes of RS distribution profiles in a part made from Inconel 718 alloy after dry machining and using three variants of cooling method are shown in Fig. 14.8.

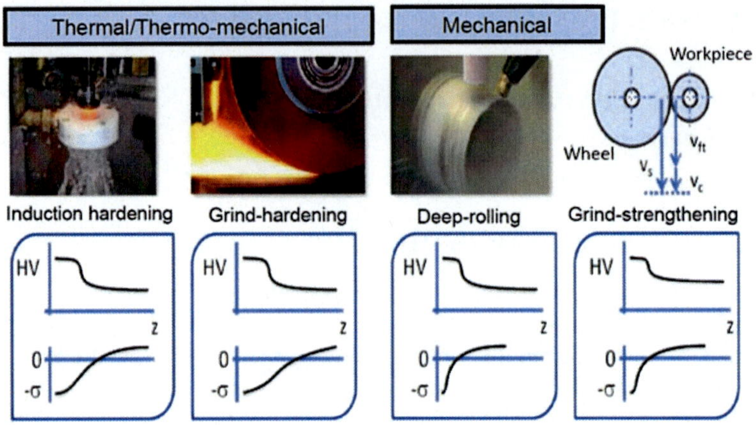

Fig. 14.7 Comparison of different hybrid processes with similar micro-hardness and residual stress profiles [10]

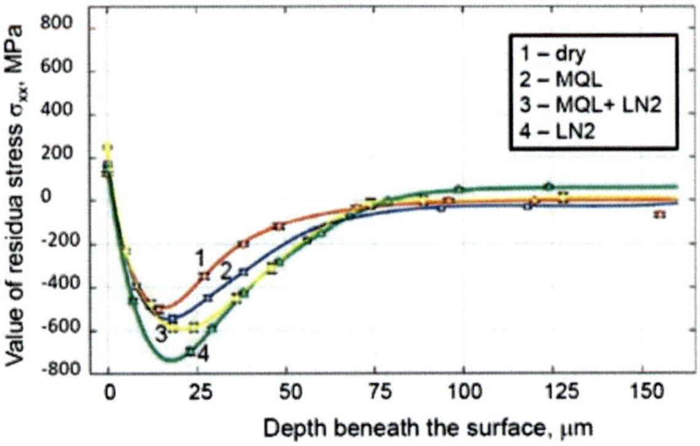

Fig. 14.8 Comparison of residual stress profiles determined in the subsurface layer after turning operations of Inconel 718 alloy using different cooling strategies; machining parameters: v_c = 60 m/min, $f = 0.05$ mm/rev [3]

In all cases considered, the hook-type RS profiles are determined but for cryogenic cooling (LN2) the highest maximum compressive stress of about 700 MPa and 20 μm beneath the surface was obtained. The intensive cooling effect leads to the generation of RSs according to the mechanical model.

In the case of laser assisted machining with a predominant thermal mechanism, as for instance in *laser assisted milling* (LAMILL) the material is subjected to complex thermal–mechanical inner loads including laser heating, heat generation during the process and mechanical loads. This leads to complex mechanism of RSs generation (RS1), which is the combination of RS2 and RS3 stresses (Fig. 14.9). On the other hand, it can be assumed that during LAMILL process the residual stresses resulting from laser scanning (LS) are superimposed with the compressive residual stresses

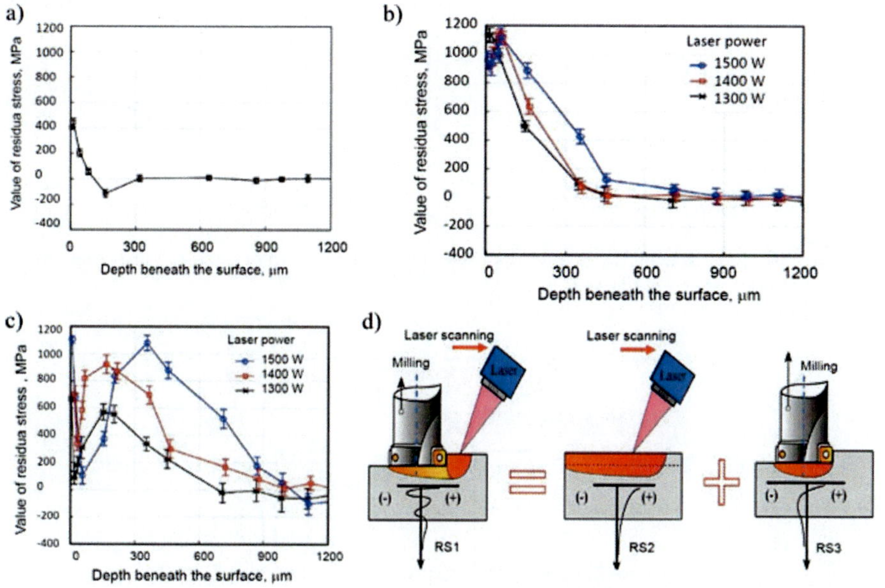

Fig. 14.9 Distribution of axial residual stresses after conventional milling (**a**), laser pre-heating (**b**), laser-assisted milling (LAMILL) (**c**) and a scheme of RS decomposition (**d**). Workpiece material-Inconel 718 [14]

generated in conventional milling (CMILL) [14]. The principle of the superposition of residual stresses generated separately in laser scanning (laser pre-heating) and in milling operation is depicted in Fig. 14.9d. For comparison, relevant distribution of RSs obtained experimentally for the three case studies are presented in Fig. 14.9a, b and c. It is worth mentioning that close to the surface (at the depth less than 100 μm) the compressive stresses generated in the milling process (Fig. 14.9a) are too low to compensate substantially higher tensile stresses generated by laser pre-heating (Fig. 14.9b). The final effect is that a complex stress state occurs with the maximum stress values close to the surface and at the distance of 150–300 μm. It should be noted that values of RSs depend on the laser power, i.e. the intensity of heat source.

Examples of RSs generated by the dominant thermal mechanism are presented in Fig. 14.10. They are generated by heating resulting from electric discharges in EDM process and pulsed laser beam (PLP) as in Fig. 14.10a, and grind-hardening and induction hardening as illustrated in Fig. 14.10b. It can be seen in Fig. 14.10a that heating by pulsated laser results in inducing very high tensile stresses of about 900 MPa. The tensile stresses are also induced to the subsurface layer after EDM roughing operation but with a distinctly lower value of about 400 MPa. On the other hand, grind-hardening and induction hardening induced longitudinal (σ_{xx}) compressive stresses of about −600 MPa localized at the depth of about 50 μm from the surface.

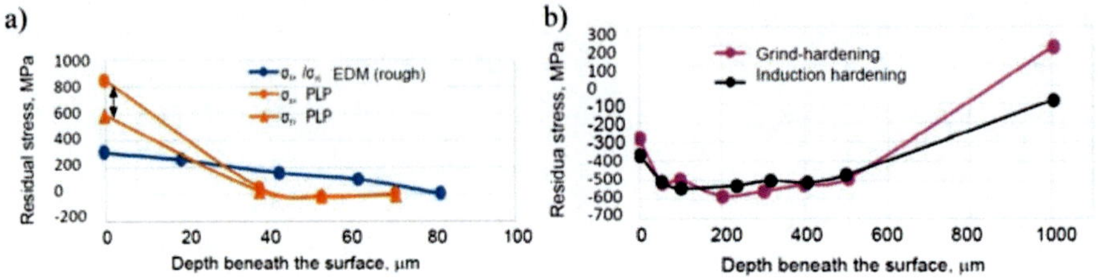

Fig. 14.10 Comparison of RS distribution after EDM and PLP processes (**a**) and after grind-hardening and induction hardening (**b**) [15]

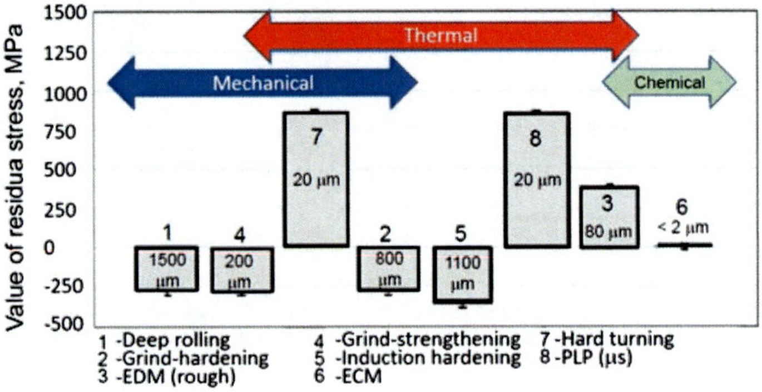

Fig. 14.11 Comparison of RS values induced to the subsurface layer after various machining processes with different participation of constituent material removal mechanisms (M-T-Ch) [15]

A comprehensive mapping of values and depths of residual stresses generated by different mechanisms is presented in Fig. 14.11. It should be noted that although the same generation mechanisms occur, the RSs penetrate at the different depth beneath the surface. One unique case is ECM process because the initial state of RSs does not change during the chemical dissolution of the material. General conclusion is that the value and sign (tensile/compressive) of RS depend on the fraction of mechanical, thermal and chemical mechanisms in the performed processes.

14.3.3 *Strain-Hardening Effect in the Subsurface Layer*

It should be mentioned that strain-hardening phenomenon is a direct effect of the penetration of intensive plastic deformation into the machined material below the cutting line.

For both surface hardening by rolling and grind-hardening the increase of material hardness is observed in a thin surficial layer with a defined thickness. In general, strain-hardening effect is typical in cutting and abrasive operations performed on ductile materials with a high plasticity index. The two hybrid machining processes

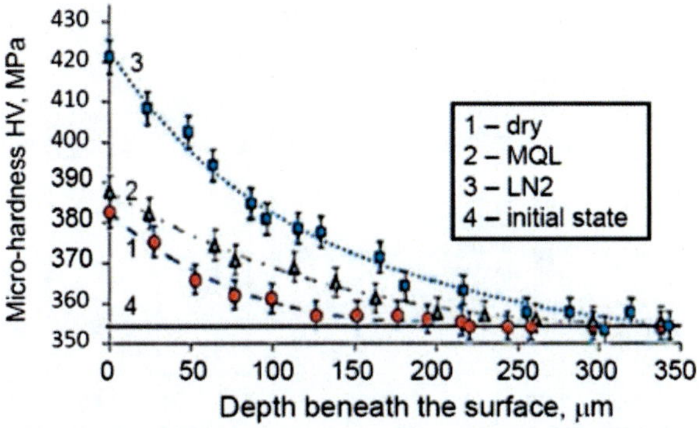

Fig. 14.12 Comparison of micro-hardness distribution after turning of Ti6Al4V titanium alloy using different cooling strategies; cutting parameters: $v_c = 150$ m/min, $f = 0.1$ mm/rev [3]

which reduce the hardness increase by strain-hardening are thermally-assisted processes utilizing thermal softening of the heated material and cryogenic machining during which material becomes harder and more brittle, and plastic deformation and corresponding strain-hardening is stopped. It is worth noticing that strain-hardening is slightly reduced in MQL assisted machining when friction is also reduced due to better lubrication of the tool-chip contact zone (plastic deformation in the SDZ is distinctly lower) [1].

Changes of micro-hardness distribution after turning of Ti6Al4V titanium alloy using dry and different cooling strategies are shown in Fig. 14.12. It can be seen in Fig. 14.12 that the machined surface generated with the assistance cryogenic cooling is harder by about 20% in comparison to the hardness of 354 $HV_{0.05}$ measured for the bulk material. On the other hand, the hardness values measured after dry and MQL assisted machining are visibly lower due to the occurrence of thermal softening effect.

Comparison of the hardness distribution resulting from strain-hardening based processes with dominant thermal mechanism is presented in Fig. 14.13. In the case of grind-hardening (Fig. 14.13a) practically a uniform distribution with the maximum HM value of about 6500 MPa was determined. In contrast, after induction hardening the hardness profile with a lower surface hardness of about 2500 MPa and a higher value of about 5500 MPa at the depth of about 10 μm was detected.

Consequently, Fig. 14.13b presents the comparison of hardness profiles after processes with a very intensive thermal mechanism, i.e. after pulsed (μs) laser processing (PLP) and roughing EDM operation. Laser pre-heating causes that the maximum hardness value of about 5000 MPa is localized at the depth of about 10 μm and in the zone beneath 15–35 μm it decreases down to the bulk material hardness. The hardness distribution profile after roughing EDM is similar with the maximum value of about 5000 MPa but with a rapid decrease at the lower depth of about 15 μm. Probably, this effect results from corresponding microstructural change by material re-quenching observed in the zone of 15–35 μm below the machined surface (see also Fig. 14.14).

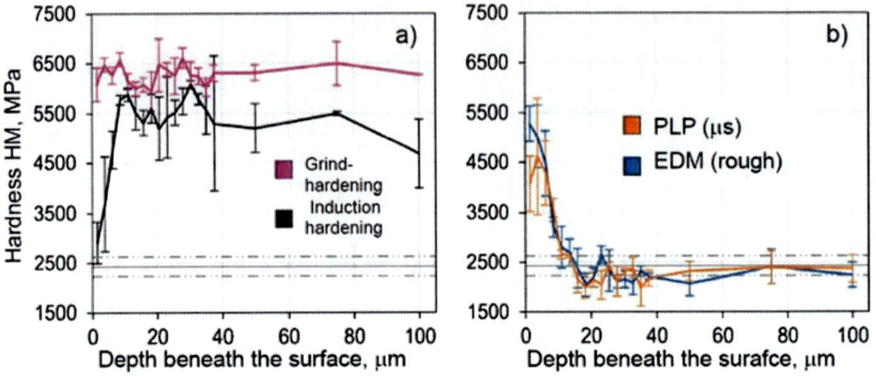

Fig. 14.13 Hardness distributions in subsurface layer after grind and induction hardening (**a**) pulsed laser pre-heating and EDM (**b**) [9]

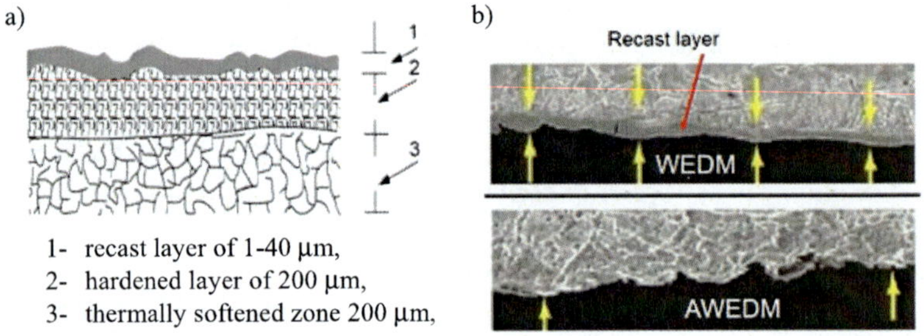

Fig. 14.14 Structure of the heat affected zone (HAZ) after EDM process (**a**) and removing re-melted layer in AWEDM process (**b**) [3, 17]

14.3.4 Changes of Material Microstructure and Surficial Defects

Changes in the material microstructure and surface defects in the form of microracks are particularly possible in machining processes with a predominant thermal mechanism and extremely high temperature gradients in the surficial layers [4, 9, 16]. In EDM processes discharges in the plasma channel reach the temperature value of 8000-12000^{0}C and cause the metallurgical alterations in the subsurafce layer (Fig. 14.14a). Simultaneously, an amorphic recast layer of 1 µm thickness is formed for the heat source power of 5 µJ and its thickness increases up to 25 µm when a more effective heat source is applied [17]. The altered subsurafce layer consists of re-quenched and quenched layers, the heat affected zone (HAZ) and precipitations on the grain boundaries. In general, the altered subsurafce layer penetrates at the depth of 0.127 and 0.075 mm after roughing and finishing EDM operations respectively. From this point of view, hybrid processes with EDM which reduce the width of the HAZ are particularly important for industrial practice (see Chap. 9). Figure 14.14b

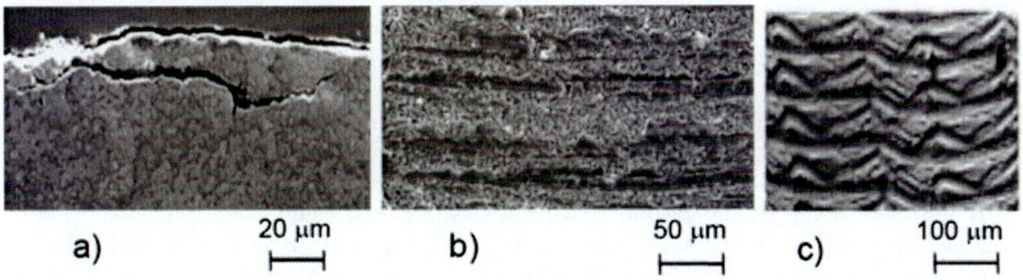

Fig. 14.15 Examples of negative influences of US vibration in USAG of ceramic material: microcracks (**a**), troughs (**b**) and chatter marks reproduced on the machined surface (**c**) [3, 6]

shows the cross-section of the subsurface layer generated in wire-EDM (WEDM) and fixed-abrasive wire EDM (AWEDM) processes. It is evident from Fig. 14.14b that the recast layer of about 5 μm thickness is generated in WEDM process but it is eliminated when AWEDM process is employed.

Because EDM is a basic machining process in aircraft industry, e.g. for producing cooling channels and sealing grooves it is necessary to incorporate additional etching to remove the recast layer. New hybrid processes use generators of ultra-short electric impulses (impulse time below μs) with high frequency (in the range of MHz) [3].

However, also negative effect can be achieved when using hybrid removal processes. The example is US vibration assisted grinding process (RUAG-*rotary ultrasonic assisted grinding*) for producing deep cavities in parts made from zirconia (ZrO_2) stabilized oxide (Al_2O_3) ceramics [6]. Figure 14.15 shows some frequently appearing surficial defects including thermal microcracks (a), troughs (b) and surface texture with visibly reproduced chatter marks. For these reasons it is not recommended to use the USAG hybrid process to machine miniature fragments in ceramic parts. Surface defects also limit the application of laser-assisted processes for ceramic parts, e.g. in micro-grinding of sintered reaction bonded SiC (SRB-SiC) ceramics outlined in Sect. 9.3.2.

References

1. Grzesik (2017) Advanced machining processes of metallic materials. Elsevier, Amsterdam
2. Grzesik W (2016) Prediction of the functional performance of machined components based on surafce topography: state of the art. J Mater Eng Perfor 25:4460–4468
3. Jawahir IS, Brinksmeier E, M'Saoubi R, Aspinwall DK, Outeiro JC, Meyer D, Umbrello D, Jayal D (2011) Surface integrity in material removal processes: recent advances. CIRP Ann Manuf Technol 60(2):603–626
4. Grzesik W (2016) Influence of Surface texture produced by finishing operations on their functional properties. J Mach Eng 16(1):15–23
5. M'Saoubi R, Chandrasakaran H, Outeiro JC, Dillon OW, Jawahir IS (2008) A review of surface integrity in machining and its impact on functional performance and life of machined products. Int J Sust Manuf 1(1–2):203–236
6. Lauwers B (2011) Surface integrity in hybrid machining processes. Proc Eng 19:241–251

7. Wang B, Shangqiang L (2018) Influence of tool structure, tool material and tool wear on machined surface integrity during turning and milling of titanium and nickel alloys- a review. Int J Adv Manuf Technol 98(5–8):1925–1975
8. Brinksmeier E, Gläbe R, Klocke F, Lucca DA (2011) Process signatures – an alternative approach to predicting functional workpiece properties. Proc Eng 19:44–52
9. Grzesik W, Kruszyński B, Ruszaj A (2010) Surface integrity of machined surfaces. In: Davim JP (ed) Surface integrity in machining (Chap. 5). Springer, London
10. Lauwers B, Klocke F, Klink A, Tekkaya AE, Neugebauer R, Mcintosch D (2014) Hybrid processes in manufacturing. Ann CIRP Manuf Technol 63(2):561–583
11. Ha J-H, Lee C-M (2019) A study on the thermal effect by multi heat sources and machining characteristics of laser and induction assisted milling. Mater 12:1032–1044
12. Zhai Ch, Xu J, Li Y, Hou Y, Yuan S, Liu Q, Wang X (2020) The study on surface integrity on laser-assisted turning of SiCp/2024Al. Int J Optomech 14(1):29–43
13. Przestacki D (2016) Laser-assisted machining of difficult to cut materials. In: Diamond and carbon composites and nanocomposites (Chap. 4). Intech Open, pp 89–106. https://doi.org/10.5772/63417
14. Xu D, Liao Z, Axinte D, Sarasua JA, M'Saoubi R, Wretland A (2020) Investigation of surface integrity in laser-assisted machining of nickel based superalloy. Mater Design 194: 108851
15. Borchers F, Clausen B, Eckert S, Ehle E, Epp J et al (2020) Comparison of different manufacturing processes of AISI 4149 steel with regard to surface modification and its influencing depth. Metals 10:895
16. Brinskmeier E, Klocke F, Lucca DA, Sölter J, Meyer D (2014) Process signatures – a new approach to solve the inverse surface integrity problem in machining processes. Proc CIRP 13:429–434
17. El-Hofy H (2005) Advanced manufacturing processes. McGraw-Hill, London

Chikhristu ndi Chilengedwe

Bukuli lafalitsidwa ndi
Kachere Series,
P.O. Box 1037, Zomba, Malawi
ISBN 99908-76-03-7
(Buku la Mvunguti no. 10)

Chithunzi cha pamwamba pa buku chinajambulidwa ndi a Chilwa Wetland and Catchment Area Management Project

Maonekedwe a buku: Mercy Chilunga
Lapulintidwa m'Malawi momuno ndi a Montfort Media, P.O. Box 280, Balaka